PROJECT MANAGEMENT
A TECHNICIAN'S GUIDE

PROJECT MANAGEMENT
A TECHNICIAN'S GUIDE

Leo Staples

ISA TECHNICIAN SERIES

Southern Wake Campus Library
Wake Tech Community College
9101 Fayetteville Road
Raleigh, NC 27603

Notice

The information presented in this publication is for the general education of the reader. Because neither the author nor the publisher has any control over the use of the information by the reader, both the author and the publisher disclaim any and all liability of any kind arising out of such use. The reader is expected to exercise sound and professional judgment in using any of the information presented in a particular application.

Additionally, neither the author nor the publisher has investigated or considered the effect of any patents on the ability of the reader to use any of the information in a particular application. The reader is responsible for reviewing any possible patents that may affect any particular use of the information presented.

Any references to commercial products in the work are sited as examples only. Neither the author nor the publisher endorses any referenced commercial product. Any trademarks or tradenames referenced belong to the respective owner of the mark or name. Neither the author nor the publisher makes any representation regarding the availability of any referenced commercial product at any time. The manufacturer's instructions on the use of any commercial product must be followed at all times, even if in conflict with the information in this publication.

Copyright © 2010 International Society of Automation

All rights reserved.

Printed in the United States of America

10 9 8 7 6 5 4 3 2

ISBN: 978-1-934394-75-5

No part of this work may be reproduced, stored in a retrieval system, or transmitted in any form or by any means, electronic, mechanical, photocopying, recording or otherwise, without the prior written permission of the publisher.

ISA
67 Alexander Drive, P.O. Box 12277
Research Triangle Park, NC 27709
www.isa.org

Library of Congress Cataloging-in-Publication Data

Staples, Leo.
  Project management : a technician guide / Leo Staples.
     p. cm. -- (ISA technician series)
  ISBN 978-1-934394-75-5 (pbk.)
  1. Project management. I. Title.
  T56.8.S73 2010
  658.4'04--dc22
                                            2010005982

DEDICATION

This book is dedicated to
Ray Cahill, Oscar Selling, and Cletus Simunek whose support,
guidance, and encouragement have been invaluable to my career.

ACKNOWLEDGMENTS

I want to thank my parents, who moved my two brothers and me (city boys) to the farm. There they shared with us their faith, their love for this beautiful world that God created, and they instilled in us a strong work ethic. These principles have served me well in my life and in my career. Thank you Mom and Dad!

Special thanks to my wife, Julie, for the love, support, and encouragement that she has given me all these years. As with so many of my ISA endeavors, the completion of this book put extra duties on her.

Thanks also to my daughter, Andrea, and my son, Jonathan, who realize their father is not perfect, but still love him.

I want to thank my company OGE Energy Corp (OG+E) for all the opportunities I have been given. I consider myself very blessed to work for a company that believes that we have a responsibility to develop leaders for the company and to provide leadership to organizations outside of OG+E.

Finally, I want to thank the members, staff, and volunteer leaders of the International Society of Automation (ISA). Since I joined ISA in 1989, I have been encouraged to take on challenges that I might have considered beyond my reach. This book is yet another example.

TABLE OF CONTENTS

Chapter 1 Project Management Overview 1

Chapter 2 Project Documentation Requirements 5

    2.0   Site Procedures 5

    2.1   Safety Procedures 6

            2.1.1   Control of Hazardous Energy. 8

            2.1.2   Work Authorization Permit 8

            2.1.3   Material Safety Data Sheets 8

    2.2   Physical Security Procedures 10

    2.3   Project Engineering Documentation 12

    2.4   Documentation Location & Media Types 14

    Summary. 15

    References 16

    Review Questions 16

Chapter 3 Communication Requirements 19

    3.0   Communications Plan 19

    3.1   Basic Communications 21

            3.1.1   Interpersonal Communication Skills 21

            3.1.2   Writing Skills 22

            3.1.3   Oral Skills. 24

            3.1.4   Facilitation Skills 25

    3.2   Communication Techniques. .26

    3.3   Other Communications Concerns 28

            3.3.1   Corporate Politics. .28

            3.3.2   Terminology 28

		3.3.3	PMT Diversity Issues 29

Summary. 30

References 31

Review Questions 31

Chapter 4 Project Planning & Resource Coordination 33

	4.0	Basic Duties Of Other Disciplines 33
	4.1	Project Planning. 35
	4.2	Task Assignments 38
	4.3	Overall Project Coordination 38
	4.4	Individual Responsibilities 38
	4.5	Supervision 39
	4.6	Time Management. 39

Summary. 40

Review Questions 41

Chapter 5 Project Monitoring & Control. 43

	5.0	Monitoring Methods 43
	5.1	Controls 44
		5.1.1 Change Management 45
	5.2	Reporting 46

Summary. 47

Review Questions 47

Chapter 6 Project Administration 49

	6.0	Document Management. 49
	6.1.	Document Organization. 51
		6.1.1 Document Review 52
		6.1.2 Document Approval Process 53
		6.1.3 Communications 53

Table of Contents

6.2	Equipment & Personnel Records	54
6.3	Print Management	54
6.4	Archiving Project Records	56
Summary		57
Review Questions		58

Chapter 7 Project Closeout **61**

7.0	Termination	62
7.1	Transfer of Ownership	63
7.1.1	Responsibilities	63
7.1.2	Documents	63
7.1.3	Equipment and Materials	64
7.1.4	Facilities	64
7.1.5	Personnel	65
7.2	Lessons Learned Session	66
7.3	Project Closeout Report	67
Summary		67
Review Questions		68

Chapter 8 An Automation Project **71**

8.0	Phase I – Initiation	71
8.1	Design	72
8.2	Phase II – Planning	72
8.2.1	Vendor Selection	73
8.3	Phase III – Execution	74
8.3.1	Factory Acceptance Test	74
8.3.2	Installation	75
8.3.3	Startup	75
8.4	Phase IV – Closeout	76
8.5	Summary	78

Figures 79

Acronyms 83

Answers to Review Questions 85

Index 97

ABOUT THE AUTHOR

Leo Staples has more than thirty years of power generation industry experience, working in the Power Supply Division of OGE Energy Corp., and has pursued post-secondary studies in project management. A few noteworthy projects include:

- Project Manager for a project to design and produce a fully computerized boiler optimization test system for fossil fuel power generation units
- Technical Lead for a Continuous Emissions Monitoring project, which consisted of 13 power generation units
- Project Manager for the electrical, instrumentation, and controls portions of a project to reactivate two 1940s fossil fuel power generation units

An ISA Fellow, Leo has authored numerous papers, given technical presentations at ISA Technical Conferences and Division Symposia, and served as a reviewer for ISA technical papers and books.

From 1994 through 2008, he managed a large technical workgroup that supported OGE Energy Corp.'s power plants in the areas of automation, testing and reliability. He currently serves as the Compliance Management Leader for the Power Supply Division.

PREFACE

My experience in project management started in the automation profession as an I&C technician in the late '70s. Early in my career much of the project work centered on converting pneumatic transmitters and indicators to analog electronics (level, flow, pressure, and temperature). Looking back, I see that I actually managed projects before I even knew there was such a thing as project management.

My first involvement with a Project Management Team (PMT) was a project to replace a power plant burner management system. The PMT members included staff from corporate design engineering and a plant engineer. My role was to support data requests by the PMT. The PMT showed up on site, toured the existing system, and obtained a set of "as-built" prints on the unit. During the site visit, the PMT made it clear that they did not require any input from the facility craft positions (electrical, technical, mechanical, and operations). They were quite surprised when they returned for installation and found that several items on the "as-built" prints were no longer part of the system. The failure by the PMT to include site personnel on the team and take advantage of their local knowledge was a lesson learned which I never forgot.

In time, I moved from supporting the efforts of the PMT to serving in positions of leadership (project technical lead and project manager). To date my project management experience includes the following areas:

- Measurement, analysis, control
- Boiler Optimization Test system design
- Natural Gas Measurement (orifice, ultrasonic, and turbine members)

- Continuous Emissions Monitoring Systems
- EPA RATA Test system design
- Compliance (regulation and programs)

Along the way, I have been blessed to work with extremely talented people who were always willing to share their project management skills. In part, I agreed to write this book as a way to honor those individuals who mentored me throughout my career.

1
PROJECT MANAGEMENT OVERVIEW

Automation projects are one of the most challenging and rewarding aspects of the job of a Control System Technician (CST). In order to meet these challenges, a CST must understand project management and the CST's role on the project management team. This book defines a CST as any technician working on an automation project or one who is working toward certification.

Industry continues to deploy automation technologies, which provide improved abilities to share information and improve performance. The rapid advance technology does present unique challenges for the technicians, technologies, and engineers involved in project management. In his book, *Successful Instrumentation and Control Systems Design* (2004), Michael Whitt makes the case that order in a project doesn't just happen, rather it must be imposed. The need to bring order to the projects led to the development of project management.

So what is project management? Project management is managing and directing people, time, other resources, and costs for a predefined time period to achieve project goals and objectives (Cockrell, 2001). Automation projects range from something as simple as replacing a set of field devices (e.g., installation of digital transmitters to replace analog electronic transmitters) to as complex as the construction of a new facility.

And what is the role of a CST in a project? A CST's role varies by industry and by the nature and scope of the project. A senior control system technician may serve as the Technical Lead or even Project Manager for a project; while the role of an entry-level technician is usually limited to providing technical support to the Project Management Team (PMT). In this role a CST may or may not be considered a member of the PMT.

The role of a CST varies by factors that include:

- The complexity of the project
- The technician's level of experience
- The technician's level of education/certification
- The technician's knowledge and skills in project organization and project administration.

In 1995, the International Society of Automation (ISA) created the CCST® (Certified Control Systems Technician®) certification to provide third-party recognition of a technician's knowledge and skills in industrial automation. Candidates must meet minimum requirements for work experience and education and must successfully complete the certification test. The examination includes the following major performance domains:

I. Calibration
II. Loop Checking
III. Troubleshooting
IV. Start-Up
V. Maintenance/Repair
VI. Project Organization
VII. Administration

Due to the technical nature of key graphics, color versions are shown in the figures section on pages 79-82.

Books on project management typically focus on the theory of project management and the role of the project manager or project engineer. This text explores project management from a technician's perspective. Subjects include documentation requirements, communication requirements, planning & resource coordination, monitoring & control, administration, and closeout. These subjects are covered in domains six (project organization) and seven (administration) of the ISA Certified Control Systems Technician (CCST) exam. While the text does focus on Domains VI and VII, readers will learn the roll all seven domains play in the various stages of project management.

REFERENCES

1. Whitt, Michael D., *Successful Instrumentation and Control Systems Design* (International Society of Automation, 2004), 3.

2. Cockrell, Gerald W., *Practical Project Management: Learning to Manage the Professional* (International Society of Automation, 2001), 3-4.

2
PROJECT DOCUMENTATION REQUIREMENTS

The documentation requirements for a given project include all materials used in design, planning, construction, operation, and maintenance. Each year industry faces increased pressures from regulatory agencies to maintain a safe and secure workplace. In the United States, these agencies include the Occupational Safety & Health Administration (OSHA), the U.S. Environmental Protection Agency (EPA), State Departments of Environmental Quality (DEQ), the U.S. Department of Transportation (DOT), and the U.S. Department of Homeland Security (DHS). Industry has developed various procedures in order to comply with these requirements, as well as those imposed by state and local governments. As a result, project documentation now extends far beyond traditional prints and drawings.

2.0 SITE PROCEDURES

Each site presents a complex set of safety, security, operations, and environmental issues. These vary greatly from industry to industry and are often unique to a specific location. To address them, most companies maintain site procedures for each location (see example in Figure 2-1).

FIGURE 2-1
Typical Site Procedure

Instructions & Methods
Revision Date   07/07/2007
Prior Revision Date 04/02/2006
Site & Safety Standards Guidelines
This Instruction and Method applies to all Company members, contract personnel, governmental agency personnel, and visitors who enter or perform any services on property or sites owned or operated by the Company.

Site procedures establish the operating authority for the facility. Typically, the site operation authority is responsible for maintaining and administering the site procedures.

Site procedures also define the work that can only be performed by the operating authority. Examples include the issuing of clearances and work permits. While all industries face certain requirements, major differences do exist. It is important to note that even plants of similar design can present unique challenges. The site procedures are an important tool for the PMT since all employees (company and contract) are responsible for acquiring a thorough understanding of the site procedures for the specific facility.

2.1 SAFETY PROCEDURES

All personnel are responsible for understanding the hazards that may be present and the type of Personal Protective Equipment (PPE) that is required. Companies are required to develop written procedures for personnel, whether employees or contractors, to safely carry out job duties.

Certain hazards exist at all sites. These include, but are not limited to:

- High pressures and temperatures (steam, water, oil)
- Combustible gases
- Chemicals
- High voltage electrical equipment
- Rotating equipment
- Confined spaces entry hazards
- Breathing hazards
- Noise
- Trip and fall hazards

Safety is an important aspect of project management. PMTs may be comprised of people who do not regularly work together. Often, some or all of the team members are not familiar with the facility.

While most accidents are preventable, they do happen. In an effort to maintain a safe workplace and minimize the effects of accidents (injuries, equipment damage, and lost production), most companies develop an Emergency Response Plan (ERP). Typically, the corporate Environmental Health & Safety Department develops the plan with input from the site operating authority. The ERP (Figure 2-2) covers emergencies, disasters, and accidents at a specific site.

FIGURE 2-2
Typical Site ERP Manual

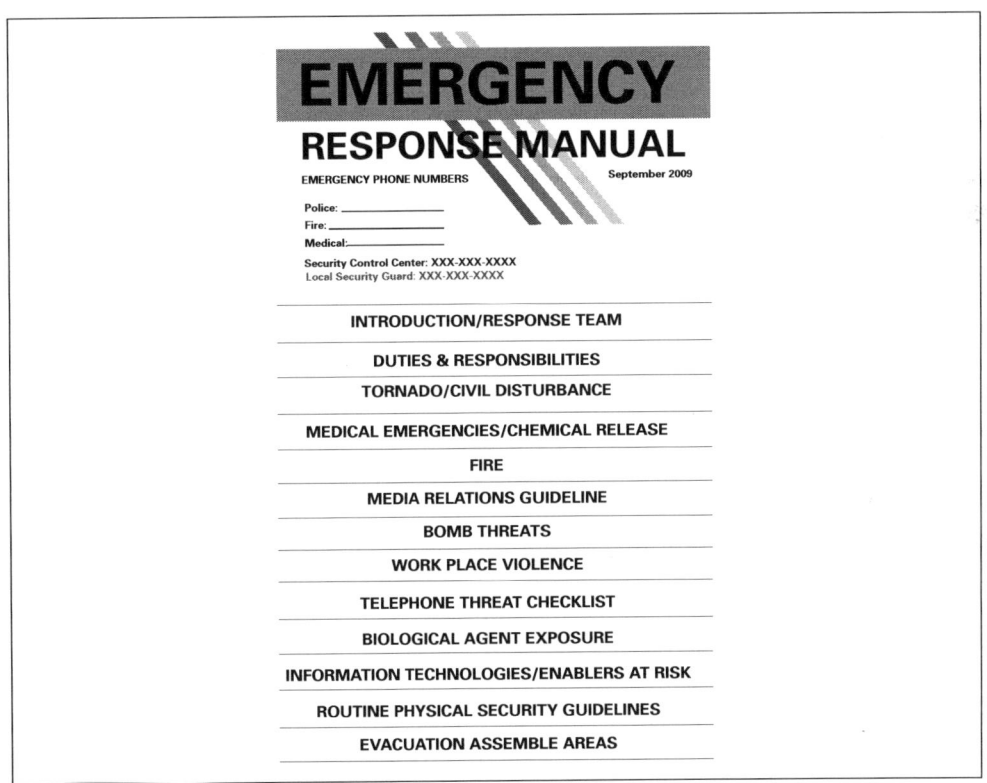

Regardless of their level of experience, CSTs must be familiar with applicable safety procedures and take the necessary steps to protect themselves. Knowing the hazards of the location, the required PPE, and the methods for responding to an emergency are the best way to stay safe.

Safety is important on and off the job. More information on this topic is available in the second chapter of the ISA book titled "*Start-Up: A Technician's Guide*" (Harris, 2000).

2.1.1 CONTROL OF HAZARDOUS ENERGY

Conditions exist in all facilities that could expose personnel to the unexpected start-up of equipment or the release of stored energy. To limit risk to personnel, the U.S. Occupational Safety & Health Administration (OSHA) developed standard 1910.147 titled, "The control of hazardous energy (lock out/tag out)." A lock out/tag out, or clearance procedure provides the means to safely isolate plant systems and individual components.

2.1.2 WORK AUTHORIZATION PERMIT

Requirements vary by industry, but in general, personnel must obtain authorization from the facility operating authority prior to starting the work. It is important to review the facility requirements of the particular facility prior to starting a project.

During a project, a CST will work under and may even hold a work authorization permit (WAP). Depending on the CST's role on the PMT, the CST may also coordinate the work of contractors. This includes ensuring that contractor personnel have obtained a WAP from the facility operating authority.

2.1.3 MATERIAL SAFETY DATA SHEETS

The OSHA Hazard Communication Standard 29 CFR 1910.1200 establishes uniform requirements for chemical manufacturers to evaluate all chemicals in order to determine hazards associated with use or exposure. The manufacturers must make this information available to the public. In the United States, the information is conveyed to users by means of container labels and material safety data sheets (MSDS). The MSDS sheet is organized into the following eight (8) sections:

- Section I - Manufacturer's Information
- Section II - Hazardous Ingredients/Identity Information
- Section III - Physical/Chemical Characteristics
- Section IV - Fire and Explosion Hazard Data
- Section V - Reactivity Data

- Section VI - Health Hazard Data
- Section VII - Precautions for Safe Handling and Use
- Section VIII - Control Measures

The international version of the MSDS is compliant with ANSI Standard Z400.1-2003. Containing sixteen sections (shown below), the international version is more comprehensive and is becoming the international norm.

- Chemical Product and Company Identification
- Composition/information on ingredients
- Hazard(s) identification
- First-aid measures
- Fire-fighting measures
- Accidental release measures
- Handling and storage
- Exposure controls/personal protection
- Physical and chemical properties
- Stability and reactivity
- Toxicological information
- Ecological information
- Disposal considerations
- Transport information
- Regulatory information
- Other information

Employers are required to educate their employees and contractors on the hazards associated with the hazardous materials they work with and ensure that information is readily available. CSTs working outside the United States should become familiar with the chemical safety programs and standards of that country.

Typically, companies develop a Hazard Communication Program to meet these requirements. The program provides specific training and communication on the following:

- Information about the hazards of chemical products used at the facility
- Location of and access to material safety data sheets (MSDS)
- Proper labeling of containers
- The appropriate use and disposal of hazardous materials

Using this information, personnel can take measures to protect themselves from these chemical hazards. In some industries, a CST routinely works with hazardous chemicals while in others a CST might have limited contact. Regardless of the industry, a CST is expected to understand and follow the site Hazard Communication Program.

2.2 PHYSICAL SECURITY PROCEDURES

Site security requirements are another major factor the PMT must consider. Security requirements vary by industry and in some cases are driven by government regulations. As an example, some industries maintain tight security requirements due to the threat of industrial espionage. In the U.S. various agencies provide oversight and maintain standards for security which is often specific to the industry. For example, the NERC Critical Infrastructure Protection (CIP) standards set physical, cyber, and sabotage security requirements for the power industry. Like the other members of the PMT, a CST should be familiar with the security requirements of the particular industry(s) they serve.

The United States Department of Homeland Security maintains a homeland security advisory system to alert industry to the threat of terrorist activities. In many cases, this will limit or restrict access of personnel to certain areas of the facilities or to cyber assets. More information is available online at www.dhs.gov/index.shtm.

Project Documentation Requirements

In addition to background checks, many companies now issue photo identification (ID) badges to anyone entering their facilities. To obtain a badge may require a worker to present two forms of government identification. With few exceptions, workers are required to wear their display where it is easily visible at all times. The PMT must take into consideration the time it takes for contractors to acquire security badges. Additional time may be required when the threat level increases.

Figure 2-3 shows a typical employee and contractor ID badge. Typically this information is coded into a visitor management system and the badge is issued. In this example, both ID badges are coded to allow the holder access to particular areas of the facilities. As shown, the ID cards would give the company member and the contract employee access to the entire facility. In many cases a contractor or visitor access is restricted access or the individual may require an escort. Additional information on the contractor's badge includes their host, which could be a member of the PMT.

FIGURE 2-3
Typical Employee & Contractor Security Badges

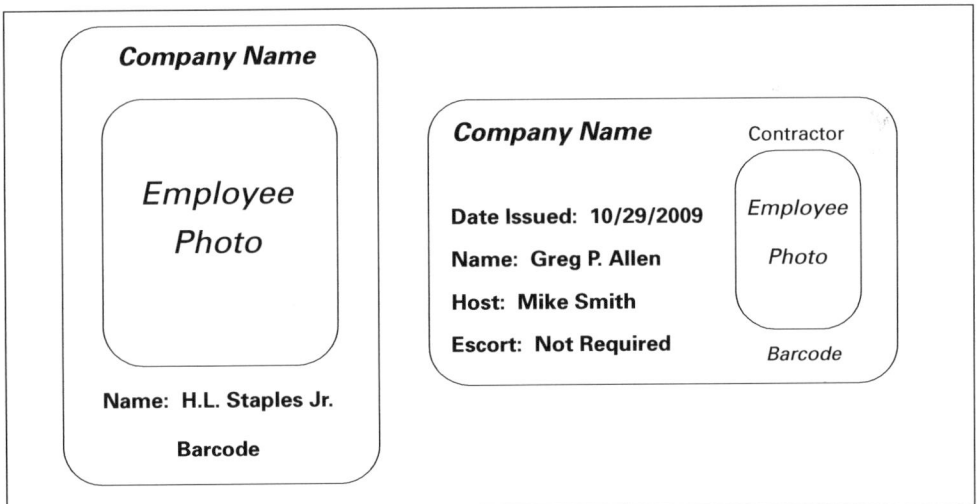

In facilities where everyone is required to badge in and badge out, the visitor management system may be programmed to provide a report on who is in the facility. This is extremely useful in emergency situations. It also provides the PMT a secondary way to verify that all workers have left the site at the end of the day.

2.3 PROJECT ENGINEERING DOCUMENTATION

Project engineering documentation includes field drawing, electronically generated prints, lists, vendor information, and other materials. Manually creating and maintaining accurate documentation is very difficult. The use of a computer-based document management system greatly improves this process.

Drawings are organized in a logical fashion in the drawing index where each item is classified according to the area of engineering (electrical/electronic, mechanical, civil, architectural, automation, etc.). As illustrated in Figure 2-4, a relationship often exists between the other engineering areas. One of the advantages of computer-aided design and drafting systems is the ability to layer drawings by engineering discipline and automatically generating a drawing index.

FIGURE 2-4
Classification of Prints & Drawings

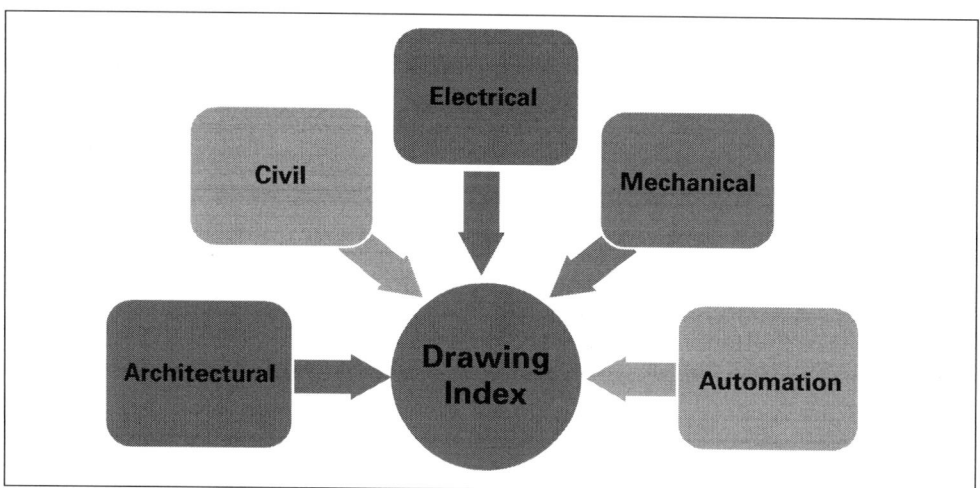

A new CST is expected to become familiar with materials relevant to the industry(s) they serve and with some level of guidance utilized materials. Likewise, an experienced CST possesses a thorough understanding of the materials you can apply the information contained within these documents.

One often-made mistake is to assume that the prints and other documents in the corporate or facility reference library are current.

Project Documentation Requirements

A CST should check the drawing index and confirm with the facility whether or not changes have been made, but not submitted for review. This may mean contacting the facility staff in that discipline to verify the drawing index number of the print in question.

For automation projects, these documents are further defined to include:

- **Process Flow Diagrams (PFDs)**, which illustrate the generalized flow of plant processes with simplified depictions of equipment and automation systems. The emphasis is on process temperatures, pressures and flows, often including heat and material balances.

- **Piping and Instrumentation Drawings (P&IDs)**, which show a much more detailed depiction of equipment arrangements and automation systems. The emphasis is on the piping, mechanical, instrumentation and automation aspects of the plant.

- **Instrument Index (List)**, spreadsheets which provide detailed information on each instrument device or function. Data includes, but is not limited to, tag numbers and instrument types, relevant P&IDs, specifications, location drawings, installation details, wiring diagrams and/or loop drawings, I/O designations and other drawings and documentation related to each instrument.

- **Specification Forms**, which list detailed information about each instrument and provide the basis for the bidding and procurement process.

- **Logic Diagrams**, which define the cause and effect relationships between instruments and the process for normal control as well as alarm and safety shut down actions.

- **Location Plans**, which show the approximate physical location of instruments as an overlay to an area plan drawing.

- **Installation Details**, which show how instruments are to be physically mounted or connected to piping systems and equipment. Also shown is the way in which electrical or pneumatic utilities are supplied to instruments, and how signal cabling or tubing is taken from the instruments.

- **Wiring diagrams and pneumatic connection diagrams**, either on a large format or on an individual loop basis (Loop Diagrams) which show point-to-point connections from the field to a local control panel or to a centralized control building or room.

- **Junction box drawings**, which show how individual signal wire pairs or pneumatic tubing lines are gathered and connected to multi-pair wiring cables or tubing bundles for transmission to control panels remote from the field.

These documents are most often developed in sequential order as a project moves from design through construction.

2.4 DOCUMENTATION LOCATION AND MEDIA TYPES

Until the advent of affordable computer technology, all project documents were printed or hand-written. Typically, companies maintained master copies in one location and working copies in another. Some companies also maintained an additional master copy at an offsite location. A CST may be required to obtain copies of these materials and determine what, if any, differences exist between working copy and master copy.

Today the majority of project documentation is developed and maintained in electronic form. Computer technology has enhanced the ease with which a PMT generates and distributes documentation. Common software systems include:

- **Business Software Suites:** Software for word processing, spreadsheet, presentation, database, email, and scheduling (project activities), packaged as a unified product.

- **Enterprise Software Suites:** Typically business-oriented tools for HR management, accounting, security, content management, customer relationship management (CRM), enterprise resource planning (ERP), business intelligence, manufacturing, maintenance, and project management.

- **Computer-Aided Design (CAD):** CAD systems are used to generate diagrams and drawings in the various engineering disciplines. For a typical automation project, a CAD system is used to produce materials for the design and construction phases and to document changes on "as-built" drawings.

It is also common for a CST to receive job assignments via an email or task list. The email or task list may direct the CST to an order generated by an enterprise software system. The order may instruct the CST to print a copy of a drawing from a CAD system. A CST needs a working knowledge of each type of document and each system being used in order to fulfill their duties on a PMT.

Managing and controlling documents is a big challenge. Many companies now use a document management system (DMS). A DMS consists of a computer system or a set of computer programs that track and store electronic and/or paper documents. As a member of the PMT, a CST may be required to obtain information from a DMS or to store information in a DMS and so must know how this is to be done.

SUMMARY

Long gone are the days of the typewriter and blueprints. While hard-copy written and print data does still exist, the majority of project management data is moving to an electronic form. Understanding the wide variety of formats and systems used to produce project management data is a challenge for a CST. This, combined with government regulations and associated compliance requirements, has also increased the complexity of an automation project. To deal with these issues, industry has developed site-specific plans that address safety, security, operations, and environmental concerns specific to the project location and that can directly affect a project. As a member of the PMT, a CST uses information in the site procedure to determine hazards associated with the site and where possible, to suggest ways to eliminate the hazard.

Obtaining accurate information is essential to running a successful project. A CST must be familiar with the various types of information used on an automation project and the systems used to develop the information and must know how to obtain this information. Throughout the project, a CST will verify the accuracy of information through field inspections and will use the appropriate system to record their findings.

REFERENCES

Occupational Safety and Health Administration. *Recommendation Format for Material Safety Data Sheets*. United States Department of Labor. July 2004. Available Online at http://www.osha.gov/dsg/hazcom/msdsformat.html.

Society for Chemical Hazard Communications. American National Standards Institute. ANSI Standardized Z400.1-2003. Dec. 2009. Available Online at http://www.schc.org/schcnewsite/pdf/revised_proposed_ANSI_400.1_std.pdf

REVIEW QUESTIONS

1. How can a CST obtain a copy of the Site Procedure for a location?

2. Explain the purpose of, and information provided by, Material Safety Data Sheets (MSDS).

3. What is the purpose and importance of lock-out/tag-out procedures?

4. What types of materials make up project engineering documentation?

5. What is the purpose of an "as-built" drawing?

6. A maintenance department typically maintains a _____ of prints for routine use.

Project Documentation Requirements

7. Logic Diagram depict _____ relationships between instruments and the process.

8. Installation details display the power and signal connections between devices, but these do not correspond to the _____ of the devices.

9. How can physical security issues affect a project?

10. A _____ _____ system that tracks and stores electronic and/or paper documents.

11. Today many _____ _____ _____ include project management software?

3

COMMUNICATION REQUIREMENTS

Effective communications are by far the biggest challenge for the Project Management Team. Poor communications increase risks associated with workplace safety, increase the amount of rework, and often lead to unacceptable results. Good communications allow the members of the PMT to coordinate their efforts and share their ideas. The PMT also uses these skills to effectively communicate with individuals and groups not directly involved in the project. Examples include the facility maintenance department, the facility operations department, and members of management.

Today, a CST must be well versed in the various forms of computer-based communication in addition to having good written and oral communication skills. In the article, "Real World Project Management: Communication" (2009), author Joseph Phillips, states that it has become increasingly important to turn our efforts toward more effective means of communication, especially since many of us are faced with more and more virtual teams operating around the globe. Mastering communication skills is becoming more complex as a CST must often take into account the language, cultural, and ethnic backgrounds of the PMT members, the contractors, and the customer.

3.0 COMMUNICATIONS PLAN

Since the success of the project depends upon clear, timely, accurate, and complete communication, it is important to establish a project communication plan. As with other aspects of project planning, the communication plan lays out the format and path of communications, and defines:

- Who is responsible for a particular communication

- Why it is required
- When it is required
- How often it is required
- What format is required
- How to handle confidential communications

A project communication plan should do the following:

- Define the project communication strategy
- Establish an agenda and date for a project kick-off meeting
- Define the roles and responsibilities of the PMT
- Define the communications process for project change management (scope of work, weather delays, workforce issues, etc.)
- Establish project status review meetings (including meeting frequency)
- List the transition plan from the PMT to the operating authority
- Establish the agenda for the Project Closeout meeting

It is important that a CST understands that project communication consists of upward, lateral, and downward communications channels. As might be expected, senior management are in the upward communications channel. Individuals or groups with direct daily involvement in project activities are in the lateral communications channel. Certainly this group includes the contractors and vendors involved in the project. It also can include corporate staff members assigned to monitor the project, the facility maintenance department, and the facility operations department. Finally, communication with PMT members occurs through the downward communications channel. By design, each of these channels addresses the communication needs of the audience. A CST must consider this as they prepare communications for a particular channel.

3.1 BASIC COMMUNICATIONS

Nearly every aspect of a project requires a CST to interact and effectively communicate with others. As shown in Figure 3-1, communication skills are grouped into four categories. Once a CST masters these basic communication skills, they should continually work to refine and add new skills through study, practice, and experience.

FIGURE 3-1
Communication Skills

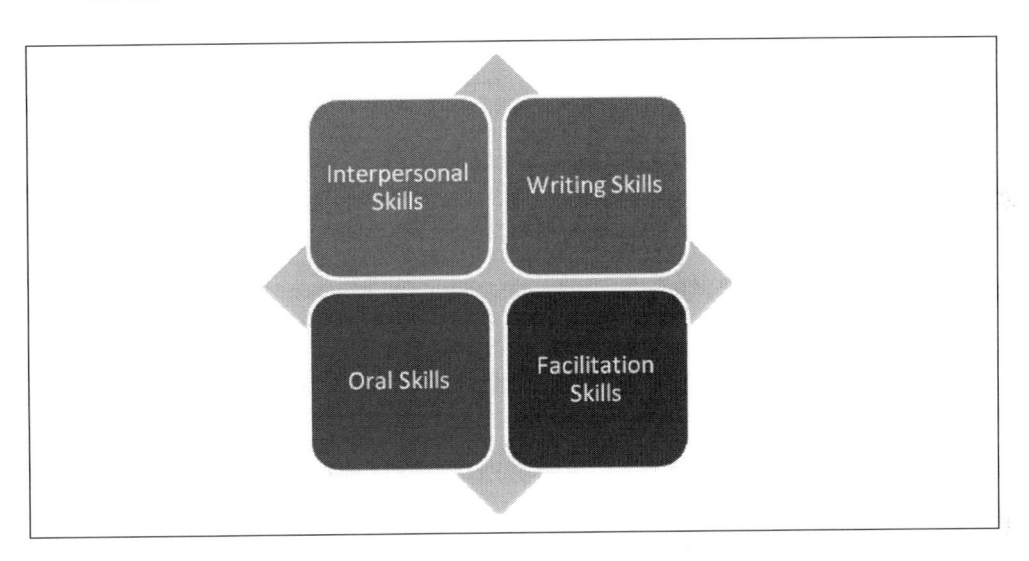

3.1.1 INTERPERSONAL COMMUNICATION SKILLS

Whether working on a PMT or performing normal job duties, a successful CST must be effective in one-on-one interaction. While most interactions are positive, the potential for interactions to become adversarial exists if they are handled inappropriately.

Fundamental interpersonal communication skills include:

1. Respond to phone calls, radio calls, emails, text messages, and other inquiries in a courteous and timely manner.

2. In face-to-face conversation, look directly at the other person when they are talking, and ask questions to ensure

mutual understanding. Focus your attention on the request or issue at hand.

3. Remember that a smile can go a long way in defusing conflict.

4. Dealing with impatient people is one of the most difficult interpersonal communication skills to master. Quite often, people simply need to vent before they can objectively view the facts of a given situation. Use good listening skills and when appropriate bring the conversation back to focus on the objectives.

5. Ensure that everyone clearly understands the situation and what steps could be taken to reach a solution.

6. Do not focus on the negative side of the situation. Let the other parties know you understand the issues and are willing to work towards a solution.

7. It is critical that difficult or insurmountable problems be communicated to the PMT.

8. Always try to identify a solution for a given problem.

Interpersonal communication skills are generally not part of educational or professional curricula. However, interpersonal communication is an essential skill, which a CST must master to be an effective member of the PMT.

3.1.2. WRITING SKILLS

As discussed in Chapter 1, a CST serving on a PMT will use writing skills to produce a variety of internal and external correspondence. While a CST is still expected to produce hand-written project communications materials, more and more PMTs are utilizing computer-generated documents. This migration to electronic media in no way reduces the need for a CST to develop good writing skills. While writing techniques vary by the type of documentation, a CST typically produces materials written from a technical perspective. In general, the following rules apply to written technical communications:

Communication Requirements

1. Always acknowledge receipt of tasks, emails, and text messages.

2. Clearly define the purpose of the message in the subject line. Keep messages as short as possible and follow an appropriate style guide for the type of document. While there are many different style guides, they all establish rules for the font size, spacing of text, documents layout, and requirements for referencing.

3. Lay out documents in an orderly fashion, using bullets and numbers when making multiple points. Reports should include headings and subheadings that guide the reader through the documents.

4. Always focus on the facts and avoid stating opinions. Effective writers avoid passive voice and using the first person in their materials. Humor is often misunderstood and can detract from the writer's intent. Always cite author and sources of any materials used.

5. Photographs, tables, graphs, and clipart can clarify points for readers.

6. Remember that materials produced on the job are the property of the company and privacy of communication does not belong to the originator.

7. Always check the names on email and text messages before responding to ensure that the response is delivered to the correct audience.

As with the other areas in project management, written communication skills will continue to evolve. A CST must monitor changes in technology and industry practices to maintain an appropriate level of writing skills.

3.1.3 ORAL SKILLS

An anonymous person once wrote, "The brain starts working the moment you are born and never stops until you stand up and speak in public." That may sound scary for those new to public speaking, but the fact remains that a CST needs good oral communication skills. As a member of a PMT, a CST is expected to be able to express ideas and make recommendations in a group setting. A senior level CST often provides a formal technical training presentation and serves as the Technical Lead on a project. To be effective, a CST must be proficient in making oral presentations to small and large groups.

As with written communication skills, it takes time and practice to become an effective oral communicator. While there are an unlimited number of books and Internet resources available on this subject, a few rules for oral presentations include:

1. A good presenter knows the material they are presenting and fits the tone, level, pace, etc., of the presentation to the audience.

2. Good presenters develop the presentation around the key points (no more than 3 to 5). If using presentation software, a good rule of thumb is to present 10 slides, talk for no more than 20 minutes, and use a large font (30 or above).

3. Good presenters have the ability to bring a subject to life for the audience. Never read a presentation; after all, the audience expects you to know the material.

4. A good presenter will practice the presentation to ensure that the oral and visual portions are coordinated.

To become an effective oral communicator takes practice. A CST should look for opportunities to make oral presentations at work and in volunteer settings.

3.1.4 FACILITATION SKILLS

An often-overlooked skill for a CST is the ability to facilitate meetings. Typically, technical members of the PMT or vendor representatives attend these meetings. The meetings occur when the group needs to share information, solve problems, make decisions, plan work, or develop buy-in. An effective facilitator understands that the people involved in these meetings often represent groups with diverse views on a situation and the objectives they wish to achieve. Understanding where each attendee stands on an issue sets the stage for a facilitator to encourage participants to share their points of view and actively participate in the discussion. Using these skills, a CST leads the meeting attendees to see value in collaborating problem solving through consensus building. These skills also serve the CST well when dealing with conflicts that arise during a meeting.

In the project management world, these meetings often come up with little or no notice. To prepare for a meeting a facilitator needs to determine the following:

- What is the purpose of the meeting?
- Who should attend?
- When and where should the meeting be held?
- What group dynamic does the facilitator need to consider?

The planning process should include the following:

- **Communications Plan:** This includes meeting notices, agendas, minutes of the meetings, and follow-up by the facilitator.
- **Ground Rules:** These help keep a meeting on track by establishing the rules for the meeting. Often a facilitator will introduce a basic set of ground rules and encourage the group to add or modify rules to meet their needs.
- **Who-Does What-By When (WWW)**: If no-one is responsible, nothing will ever get done. Using the WWW approach provides the facilitator a means to ensure that the

group agrees to an action and that someone has agreed to accept responsibility for the action. This is a great tool for following up on assignments as it includes deadlines for the action (Figure 3-2).

FIGURE 3-2
Example of a Who-Does What-By When

Who	Does What	By When
Davison	Contact the Station Operating Authority to obtain site specific information on Unit 4	1/8/2009
Leach	Contact the EH&S Department to determine what PPE is required for staff and contractors working on the Unit 4 Control Project	1/10/2009
Miller	Schedule, coordinate, and lead the Factory Acceptance Test on the Unit 4 Control System	3/15/2009
Mayberry	Produce and distribute the minutes of the meeting	1/16/2009

Regardless of their position on the PMT, a CST must be familiar with the meeting facilitation process.

3.2 COMMUNICATION TECHNIQUES

As has been mentioned, communication is an essential component of a successful project, whether it is at the interpersonal, intergroup, organizational, or external levels. Today, members of the PMT have access to a wide variety of communication techniques. In addition to knowing how to use each technique, a CST is expected to know which one is appropriate for a given situation. Figure 3-3 shows a list of current communication techniques.

Communication Requirements

FIGURE 3-3
Communication Techniques

Technique	Form	Description
Memo	Electronically produced using business suite or enterprise suite software	Formal communication used to communicate important dates, policies, and procedures
Meeting Activities Record	Electronically produced using business suite software	Formal communication of meeting notices, agendas, minutes, and WWW activities
Task List	Electronically produced using business suite software	Allows the PMT to assign and track work assignments
Email	Electronically produced using business suite software	Allows the PMT to electronically transmit text, audio, documents, and video files
Fax	Electronically produced images of documents	Allows the PMT to send copies of project documents
Instant Messaging	Electronically produced using business suite software or Provider Site on the Internet	Allows team members to communicate real-time
Telephone/video conference	Electronic or Internet based	Provides a medium to communicate in real time with the PMT, customers, and vendors who are in other locations
Project Planning	Electronically produced using business suite or enterprise suite software	Provides a medium for producing and sharing the project plan, making work assignments, managing schedules and budgets, and producing project status reports
Change Order	Electronically produced using business suite or enterprise suite software; often printed out for signatures	Provides a medium to communicate and document a change in scope to the project. Unlike many project documents, parties representing the customer and the PMT typically sign a change order

The types of communication techniques available to the PMT continually evolve, allowing the team to take advantage of new technologies. As these changes occur, a CST must develop a working knowledge of the techniques.

3.3 OTHER COMMUNICATIONS CONCERNS

If all this were not enough, the PMT faces communications issues related to corporate politics, terminology, and the diversity of the team members.

3.3.1 CORPORATE POLITICS

Corporate politics in the project management environment involves the power struggles between the various groups involved in the project. This includes disagreements between the maintenance department and the operations departments, as well as between the facility management and corporate management. Through careful observation of the interactions between the various factions, a CST can determine who the key political players are. Internal politics can and often do have a negative effect on a project. There are two major reasons a CST should know the key political players. First, this information helps a CST avoid being caught up in a political power struggle. Second, it allows the CST to take steps to ensure that all parties have confidence in the CST's work and commitment to the project goals.

3.3.2 TERMINOLOGY

The use of acronyms and abbreviation adds yet another level of complexity to project management communications. The larger the scope of a project, the more likely a CST will encounter acronyms or abbreviations that have different meanings to different disciplines. Do not fall into the trap of assuming that a common understanding exists. As an example, a CST needed a copy of the site Emergency Action Plan and sent an email to the PMT requesting a copy of the "EAP." The facility HR representative sent a copy of the Employee Assistance Program, the Environmental Coordinator sent a copy of the site Environmental Action Program, and the site Financial Analyst sent copies of the Electronic Account Payments for the project. To ensure clarity, it is best to always list the full name followed by the acronym or abbreviation the first time it appears in a communication document (Figure 3-4).

FIGURE 3-4
Proper Use of Acronyms

> In the early 1990s work by the Testing Task Force led to the development of calibration standards and intervals for power plant instrumentation. This work was documented in the book, "Testing Technicians Training" produced by General Physics Corporation (GP) and Oklahoma Gas and Electric Services (OG&E). The text set forth a uniform set of rules and procedures for standards in the areas of temperature and pressure measurement. These standards were based on the American Society of Mechanical Engineers (ASME) Performance Test Codes, the American National Standards Institute (ANSI), the National Institute of Standards and Technologies (NIST), and the International Society of Automation (ISA).

In 1943, Bell Laboratories coined the word acronym as the name for a word created from the first letters of each word in a series of words (Wikipedia, 2009). A familiar acronym is "OSHA." Today more than a million acronyms and abbreviations (which, strictly speaking, aren't spoken as words, but the distinction is often overlooked) are used by the various disciplines (Engineering, Accounting, IT, Marketing, government agencies, etc.).

3.3.3 PMT DIVERSITY ISSUES

Today, the PMTs routinely include members who are citizens of a country other than where the project is physically located. These non-domestic members may be in the country on some form of work VISA or may perform project activities while located in another part of the world. In addition to obvious logistical issues, this type of PMT faces additional obstacles to effective communication. These include:

- **Language**—While English is the common business language around the world, it is often not the first language for all personnel on a project. As an example, a CST installing thermocouples may encounter an insulating crew whose members do not all speak fluent English. For safety and quality control reasons, it is important for the CST to determine what, if any, communications barriers exist due to language.

- **Cultural**–Certainly, the opportunity for communication problems increases when the people involved come from different cultures. Developing an understanding of the social attitudes and behaviors of the people working on a project allows a CST to avoid culture-related communication problems. However, geographical differences are not the only form of cultural problems facing PMT members. Often huge cultural differences exist among the levels of an organization. These differences can also exist between the organization and the external companies (goods and services providers) working on the project. Understanding the cultural differences of the organizations involved in a project allows a CST to communicate in ways that help break down cultural barriers.

- **Politics**–The continued drive for globalization does impact the PMT as members must be aware of the political views of the country and people associated with a project. A CST working on project outside their home country is well advised to develop a basic understanding of the political views of that country. Likewise, a CST serving on a PMT with non-domestic members is well advised to develop a basic understanding of political views of these members. This knowledge is key to avoiding situations that would bring politics into the project.

Effective PMTs recognized that embracing diversity sets the stage for creativity and innovation.

SUMMARY

One constant in successful projects is effective communications. This starts with the creation of a project communications plan that is supported by and, more importantly, adhered to by everyone involved in the project. Following the plan reduces the opportunity for misunderstandings as communication paths are clearly defined and all communications are documented.

Diversity of the PMT members is one of the biggest challenges to effective project communications. It is easy to understand how a multicultural PMT could face problems related to language, cultural, and political issues. A CST working on a multicultural PMT should recognize and respect the differences within the team. The value of effective communications skills in these situations cannot be overly stressed.

During their career, a CST will experience many changes in communication as the result of new techniques, new technology, and the world's drive to globalization. Mastering the four basic communications skills is a lifelong endeavor that may well determine how far a CST progresses in their career. With these changes, and more to come, the ongoing challenge for a CST will be mastering interpersonal, written, oral, and facilitation communication skills.

REFERENCES

1. Phillips, Joseph, "Real World Project Management: Communication" Project Smart. Accessed 7, January 2009. Available from http://www.projectsmart.co.uk/real-world-project-management-communications.html

2. Wikipedia, Acronym and Initialism Definition [Online], Accessed 10 December, 2009; available from http://en.wikipedia.org/wiki/Acronym_and_initialism#cite_note-0

REVIEW QUESTIONS

1. What does a project communications plan define?

2. What are the basic communications skills that a CST must master?

3. Given that more and more project documents are produced electronically; does a CST really need good written communications skills? (Explain)

4. When is it extremely important to have mastered good interpersonal communication skills?

5. Should employees expect privacy for the communication materials (memos, text messages, email, etc.) they produce on the job?

6. Why are oral communication skills important to a CST?

7. Why does a new CST need to be familiar with the facilitation process?

8. Understanding the social attitudes and behaviors of the people working on a project allows a CST to avoid _____ _____ communication problems.

9. Why is it important for PMT members to identify an acronym or abbreviation by its full name?

4

PROJECT PLANNING & RESOURCE COORDINATION

A PMT cannot plan a project without knowledge of the resources (human, goods, and materials) needs and availability. This includes the necessary human resources (skills, abilities, staffing levels, etc.), equipment, and material. Using this information, the PMT develops a project plan which matches available resources to required tasks.

In many cases, a CST will perform research and obtain information used in the planning process. From the start of the execution phase through the closeout phase, a CST will provide feedback on the status of project tasks. This feedback is critical to resource coordination efforts.

4.0 BASIC DUTIES OF OTHER DISCIPLINES

Led by a Project Manager, the PMT consists of various professions, crafts, and trades responsible for particular aspects of the project. Depending on the size and scope of a project, these could include Engineering, Planning, Environment, Technical, crafts (electricians, mechanical maintenance, pipe fitters, welders, machinists, etc.), and Operations. PMT members may come from the customer (local and corporate), architectural & engineering (A&E) firms, construction companies, vendors, and manufacturers. In response to increased regulations and political unrest throughout the world, PMTs today may include Safety and Security staff members. Whatever their skill or affiliation, each member is responsible for performing their tasks according to the project schedule.

The size of a project often dictates the makeup of the Project Management Team. A facility may be able to support a small project with limited outside workers. On a large project, the PMT generally is made up of a combination of staff from the facility, manufacturers, and contract employees. In the case of turnkey projects, the facility staff may have little involvement in the project until the startup phase. In all cases, the PMT members are responsible for completing the tasks as outlined in the Project Plan. As an example a Project Management Team could include:

- **Project Manager (PM):** The Project Manager is responsible for all aspects of the project. These include project development, planning, communication, and reporting. The goal of the Project Manager is to ensure that the project is delivered on time, in budget, and within scope.

- **Discipline Project Engineer (DPE):** Depending on the scope and nature of the project, the DPEs can include civil, architectural, environmental, mechanical, electrical and control systems engineers (CSEs). The highly specialized CSE is responsible for all aspects of automation instrumentation and systems on a project, including planning/scoping, design, specification, installation, startup, training and eventual turnover of the new automation systems to the facility operations and maintenance personnel.

- **Technical Team Lead:** A senior CST often serves as Technical Team Lead on the PMT. This position requires in-depth technical knowledge and leadership skills. A Technical Team Lead provides expert technical guidance to the PMT, supervises technical members assigned to the PMT (facility and contract employees), and often leads the PMT when the Project Manager is not available. This is especially true on large automation projects.

- **Architectural & Engineering (A&E) Firms:** As the name implies, A&E companies are built around the architecture and engineering disciplines. In addition, A&E firms employ staff members from other areas of project management. As such, they can often provide project management from initiation to closeout. In delivering a turnkey project, the A&E firm is responsible for design, procurement, construction, commissioning, and training. A&E firms use company employees or subcontractors to support the various activities of a project.

- **Construction Companies:** Construction companies typically do not have architects and design engineers. In many cases, they specialize in a particular aspect of construction (mechanical, electrical, technical, etc.)
- **System Integration Companies (SIC):** System Integration Companies specialize in understanding the client's needs and developing systems that function together. They typically do not approve design, procurement, and project management function. In an automation project, SICs are often responsible for the control systems.

Clearly, the diversity of the PMT's membership is a challenge. Therefore, it is imperative that a CST understands the roles and responsibilities of each group and, where appropriate, of the members of that group.

4.1 PROJECT PLANNING

Phase I initiates the lifecycle of a project (Figure 4-1) with the development of an idea or concept that will meet the company's needs. Once a decision is reached to move forward on the project, a PMT is selected. Working with the customer the PMT develops estimates and the project charter, which defines the project deliverables. It is by these deliverables that the success of the project will be determined.

FIGURE 4-1
Project Lifecycle

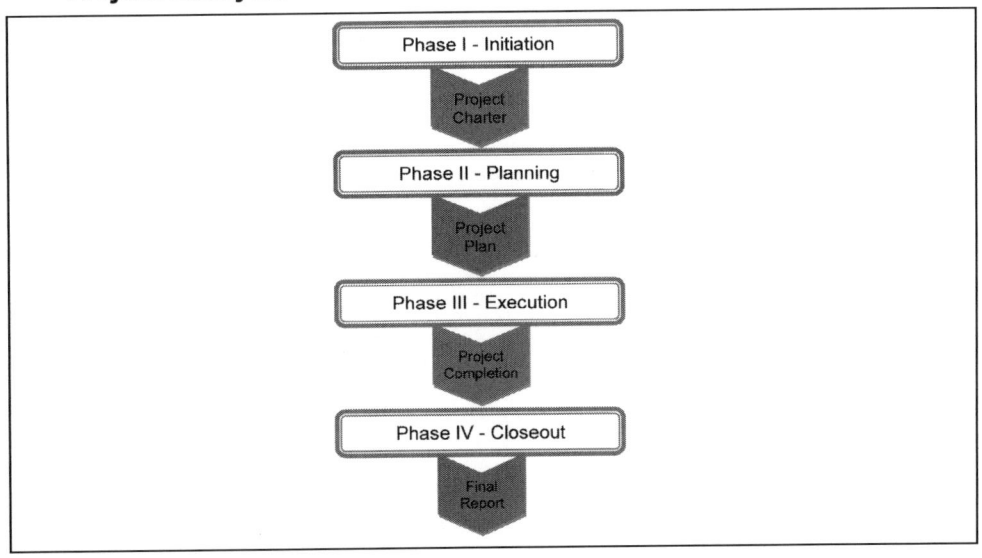

Phase I - Initiation
↓ Project Charter
Phase II - Planning
↓ Project Plan
Phase III - Execution
↓ Project Completion
Phase IV - Closeout
↓ Final Report

In Phase II, the PMT develops the project milestones and sequence of events that make the project plan and determines how the project will be completed. During this phase the PMT determines resource needs (staffing, goods, and materials), develops cost estimates, and presents the project plan to the customer for approval.

In Phase III the PMT performs the tasks, monitors the work to see that it conforms to the plan, reports progress on the project, and manages the inevitable changes that all projects face. In Phase IV, the PMT updates all project documentation (as-built drawings, instrument lists, calibration sheets, etc.), conducts a customer satisfaction survey, conducts a "Lessons Learned" review meeting, and produces a project report.

Project Planning is the development and use of schedules to plan and manage the project and produce project status reports. The various tasks necessary to complete the project make up the project scope. Once these tasks are defined, the PMT develops a work breakdown structure, which organizes the activities by discipline and by their interdependence to the other task in the project. Gantt charts and PERT (Program [or Project] Evaluation and Review Technique) charts are common methods used to organize project activities.

A Gantt chart (Figure 4-2) presents a project schedule, in a manner that highlights the dependency of one task on another.

FIGURE 4-2
Gantt Chart

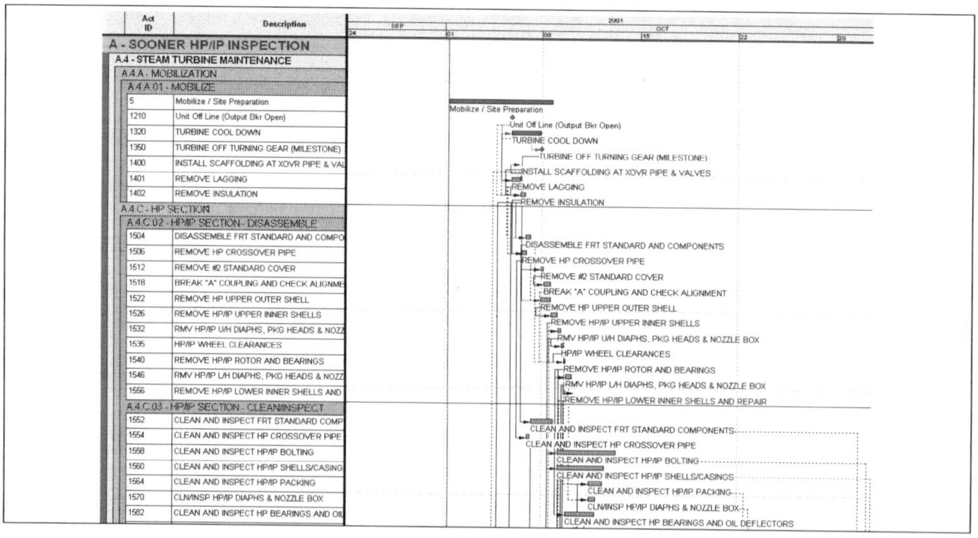

A PERT chart (Figure 4-3) is used to analyze project tasks in order to determine the time needed to complete each task and the total time to complete the project. In this example, the chart shows the project is scheduled to last six months, has eight activities (A through H) and has six milestones (10 through 60).

FIGURE 4-3
PERT Chart

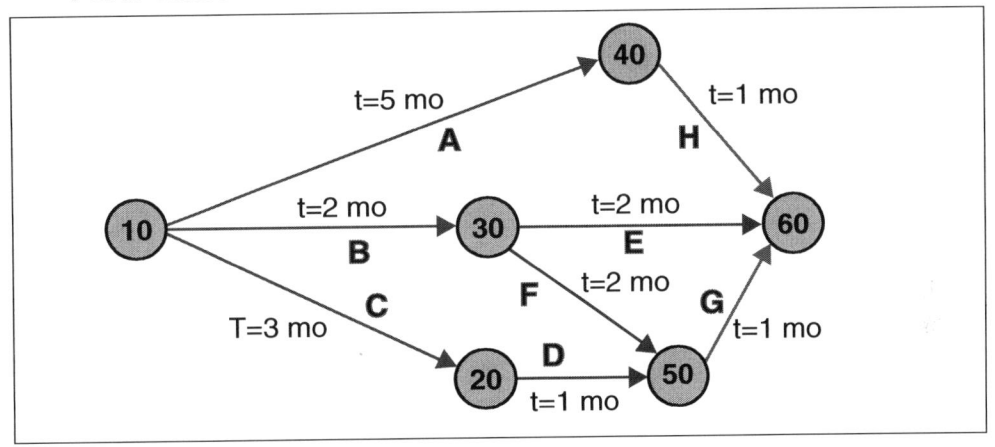

Today, companies utilize project management software to perform this work. Depending on the size of the project, the software may reside on an individual computer or on a computer network, or be a web-based application. Examples of project management software include Microsoft Project, Primavera Project Planner, and SAP Project System. By design, the software helps PMTs and their members collaborate, stay on schedule, and manage expenses.

As companies migrate to enterprise software suites, many are utilizing the project management tools that come with the suite. While more complicated, these tools integrate with the other enterprise suite applications allowing the PMT to plan, execute, and account for project activities as part of normal business procedures.

A CST must develop a working knowledge of the materials generated by a project management software package. A senior CST may require a working knowledge of the project management software in order to review the project, add tasks, or modify the project schedule.

4.2 TASK ASSIGNMENTS

As discussed in section 4.1, projects consist of interdependent tasks that can and often do disrupt the project schedule. Project management software provides an effective means of tracking task assignments and status. Tasks are assigned to individuals and to workgroups. As an example, during the design process a CST may receive a task to obtain a copy of a current Piping and Instrumentation Diagram (P&ID) of the company's control system. A CST is expected to understand how they will receive task assignments, the task work requirements, and the interdependence a particular task has with other aspects of the project. Senior CSTs serving as Technical Leads may assign tasks to members in their work area. In these situations, they are responsible for ensuring that the work is performed on schedule and to the standards established in the project plan.

4.3 OVERALL PROJECT COORDINATION

The Project Manager is responsible for leading project coordination efforts. That said, the other members of the team certainly play a critical role in the coordination process. Depending on the scope of the project, daily coordination duties may be assigned to other members of the PMT. Some PMTs use a Project Coordinator or Project Planner to perform these duties. A CST needs to understand the role project coordination plays in project management. This understanding includes providing input into the project coordination process appropriate to their position on the PMT.

4.4 INDIVIDUAL RESPONSIBILITIES

Unlike many of the members of the PMT, a CST's involvement continues throughout the project. For this reason, a CST is responsible for developing a working knowledge of the project management process. Examples of the types of project activities performed by a CST include:

- Gathering information during the initiation stage
- Providing input (knowledge of the facilities, staffing, and technical knowledge of the project equipment) during the planning stage

- Participating in the myriad of activities that take place during the execution phase
- Supporting the closeout activities

Of course, this all depends on the CST's knowledge and experience. A CST who understands the project planning process is better prepared to fulfill their assignments, which directly contribute to the success of the project. This knowledge also allows the CST to monitor project activities for potential problems that would adversely affect the project schedule.

4.5 SUPERVISION

Supervision on a PMT comes in the form of direct and indirect supervision. An example of direct supervision is the relationship of the PM to the PMT members. An example of indirect supervision is the relationship of the PMT to the staff members of a contractor. In this situation, the PMT may direct the activities of the contractor's staff members, but is not responsible for providing basic supervision to these members.

Most CSTs assigned to a project are not in a supervisory role. Still, it is important for a CST to understand the roles and responsibilities of a supervisor. For example, a senior CST serving as Technical Lead may have supervision responsibilities. The CST may be formally recognized as a supervisor or it may be understood that as Technical Lead they are in a position of leadership. To be considered for a leadership role, a CST must exhibit leadership skills. Work-related opportunities for leadership development can be limited. So how does a CST develop leadership skills? Volunteer organizations, such as ISA, provide CSTs with excellent opportunities to develop leadership skills.

4.6 TIME MANAGEMENT

Time management, as it relates to the role of a CST in project management, may seem out of place. After all, most CSTs assigned to a PMT work under direct supervision. However, time management is important to all members of the PMT as the project schedule contains a time estimate for each task. Failure to complete a task on time can have a negative effect on the project schedule. This is especially true for "critical path" items. Therefore, a CST must plan their work in order to stay on schedule. Taking a few minutes to ensure that the assignment is clearly defined and the work is planned out can make all the difference. Important considerations include:

- Location of the work
- Equipment requirements
- Support required to complete the task
- Safety/security concerns that could impact the work.

CST's with good time management skills review their assignment tasks, identify and address potential problems, and plan their work in order to meet the project schedule.

SUMMARY

The planning and coordination of projects may seem overwhelming to a new CST. Fortunately, most companies utilize a standard structure for project management. Today many companies also use certified Project Managers. Having a qualified leader who follows a standardized process greatly shortens the learning curve for new PMT members. As the CST progresses in their career, they need to develop a more thorough understanding of project planning and coordination issues. More information on project planning is available in the following ISA books:

- *Practical Project Management – Learning to Manage the Professional* by G. W. Cockrell (2001)
- *Fundamentals of Industrial Control – Project Management Strategies* by D. A. Coggan (2005)
- *A Guide to the Automation Body of Knowledge*, Second Edition by Vernon L. Trevathan, Editor (2006)

Project Planning & Resource Coordination 41

REVIEW QUESTIONS

1. What information can a CST obtain from a project Gantt chart?

2. Today many companies are migrating from stand-alone project management software to _____ _____ suites.

3. Project Management Team members can include:
    A. Customer (local site and corporate) employees
    B. A&E firms
    C. Construction companies
    D. Vendors
    E. Manufacturers
    F. All of the above

4. A Project plan is developed during Phase _____.

5. _____ involves the use of schedules to plan and manage a project and produce project status reports.

6. A _____ chart presents a project schedule, in a manner that highlights the dependency of one task upon another.

7. One advantage of enterprise suite project management applications is that the application is _____ with the other enterprise applications.

8. The project plan consists of a list of interdependent _____.

9. As a member of a Project Management Team, a _____ is expected to understand how tasks are assigned and completed work is reported.

10. For a CST, the important time management issues include:
    A. Location of the work
    B. Equipment requirements
    C. Support required to complete the task
    D. Safety/security concerns that could impact the work
    E. All of the above

5
PROJECT MONITORING & CONTROL

As discussed in Chapter 4, project planning begins when the project is initiated and ends at closeout. In this chapter, readers will learn how project monitoring and control (PMC) are used to manage the project plan. CSTs are not often responsible for the actual monitoring and control of a project. However, the information they contribute is highly valuable to this process.

Project monitoring and control techniques provide the PMT with an understanding of where the project stands in relation to the project plan, based on key indicators. These include:

1. Time: As it relates to the project schedule
2. Costs: In relation to the project budget
3. Performance: In comparison to the project design specifications

In order to manage the plan, the PMT must monitor these indicators and, when necessary, take corrective action.

5.0 MONITORING METHODS

The methods used to monitor project activities are driven by the level of technology the Project Management Team is applying. As we have seen, many companies use enterprise software suites that automate the project management monitoring process. While much more complicated to set up, they do collect all information required to monitor the fulfillment of the project plan. Typically, this involves some level of written (paper and/or digital) reporting on the status of project activities. Examples of project reports include:

- Diaries or logs

- Timekeeping reports
- Work orders
- Project change orders
- Vendor/supplier status reports
- Project status reports
- Project update meeting minutes

All members of the PMT contribute to the monitoring process through the information they submit. The extent and level of project monitoring duties for a CST are directly related to their specific project responsibilities.

Monitoring helps the PMT ensure that the appropriate resources (skills, personnel, goods and materials, funding, etc.) are available to support planned activities and that work is progressing according to the project plan. Variances from plan occur no matter how well a project is planned. Knowing the status of the various project activities allows the PMT to determine which activities have been affected by a variance. In turn, the team develops and evaluates alternative solutions to the problems that arise due to the variance.

It is relatively easy to put into practice a monitoring process for goods and materials. Monitoring the performance of people is another matter. It requires in-depth technical knowledge of the area in question and the time to monitor the work. As an example, a senior CST assigned as Technical Lead on an automation project should have the skills and knowledge required to monitor and report on the performance of contractors installing a control system.

5.1 CONTROLS

The PMT uses project monitoring methods to compare what is actually happening in the project to the project plan. Control activities include actions that ensure that project schedule, budget, and resource allocation continue to align with the project plan. As an example, the project plan includes 60 hours for technicians to install a PLC in a period from December 1 through December 4. The PMT monitors time charges to this activity and progress to completion. With this information they can:

- Determine if the work is progressing on schedule and where necessary provide additional resources
- Determine if the task will be completed within budget and where necessary reallocate funds from other areas of the project in order to stay within budget

While this sounds simple, complex projects often involve thousands of interrelated activities. The failure to complete a seemingly minor task at the scheduled time can bring a project to a halt. As we have noted, variances occur no matter how well a project is planned. When a variance occurs, the PMT must determine if the variance is acceptable in order to focus their attention on other activities that have the potential to derail the project.

The CSTs contribute to both the project monitoring and project control processes by providing information. Examples include:

- Submitting completed project work orders (scheduling control process)
- Verifying that a shipment of control system components matches the order (budget and resource allocation control process)

A senior CST may be tasked with monitoring the progress of work in a specific area and when necessary take steps to ensure that the correct level of resources is allocated to a project task.

5.1.1 CHANGE MANAGEMENT

During a project, the PMT will encounter problems that require action outside of the approved scope of the project. Changes to the project's scope occur for a variety of reasons. A few of the common reasons include:

- Design flaws
- Delivery issues
- Time constraints
- Resource shortages
- Changing customer needs

A "Change Order" is a formal process of identifying problems, evaluating options, choosing a solution, requesting approval, and implementing the change. An essential part of the change order process is ensuring that all actions are reflected in the appropriate project documentation (prints, drawings, calibration records, equipment lists, etc.).

A CST plays a variety of roles in the change order process. Their work may be as simple as identifying a range error when performing equipment calibrations and submitting a request to approve the change. For more complex problems, a CST may participate in the process of developing a solution to the problem. More information on this subject is available in the ISA book titled *"Start-Up: A Technician's Guide"* (Harris, 2000).

5.2 REPORTING

As discussed in Chapter 3, the PMT uses a variety of communication techniques to convey information and report on the status of the project. Regularly scheduled project status reports are the best way to ensure that the PMT and the stakeholders have a mutual understanding of the true state of the project. The methodology and reporting interval for the project status report are defined in the project communications plan. A typical project status report includes the following:

- Schedule: Reviews the list of activities completed during the reporting period and presents the work schedule for the upcoming period.

- Budget: Presents a comparison of the current project expenditures to the budget and projected expenses for the remaining activities.

- Risks: Reports on the status of previously identified areas of concern, identifies new areas of concern, and details plans to manage these risks.

Risk to a project includes such obstacles as delayed delivery of key materials and subsystem incompatibility. Risks may also

Project Monitoring & Control

include labor shortages, customer change orders, and delays in decision-making by the PMT or stakeholders.

SUMMARY

The process of monitoring and controlling a project starts when the project is initiated (Phase I) and ends with the project closeout report (Phase IV). The effectiveness of these processes is reflected in the PMT's project status reports. As a member of the PMT, a CST contributes to these processes in many ways. The contribution of a CST with limited technical and project experience may include completing work orders, reviewing material delivery packing lists for the automation equipment (controls, systems, devices, etc.), and collecting information for the status report. In addition to these duties, a senior CST may also provide input used to estimate future project expenses, identify risk, and plan risk mitigation. All of these could potentially affect the success of the project.

REVIEW QUESTIONS

1. What are some of the ways a CST contributes to the monitoring process?

2. _____ _____ software automates portions of the project monitoring and control process.

3. On projects that utilize enterprise software suites for planning and scheduling tasks, a CST may log into the system and confirm a _____ _____ once a task is complete.

4. Monitoring helps the PMT ensure that the appropriate _____ are available to support activities and that work is progressing according to the project plan.

5. _____ activities involve actions that ensure the project schedule, budget, and resource allocation align with the project plan.

6. When a CST validates shipping documentation against the equipment delivered, they are helping control the project _____.

7. Which of the following items are included in a project status report:

   A. Schedule

   B. Budget

   C. Risk

   D. All of the above

8. The actual methodology and reporting interval for the project status report is defined in the _____ _____ plan.

9. Material delivery delays and subsystem incompatibility are potential _____ to a project.

10. The duties of a senior CST serving as Technical Lead on a project may include providing input to help in identifying risks and developing a _____ plan.

6

PROJECT ADMINISTRATION

Under the direction of the Project Manager, the PMT performs tasks that allow the team to administer a project. The term *project administration* is used to describe the support functions of tracking, reporting, and internal communications. Performing these tasks allows the PMT to manage the resources (human and materials), financial, and logistical issues of the project. Examples include planning, time management, cost monitoring and control, billing (typically for external clients), and reporting.

This chapter highlights the overall scope of project administration, as well as the role a CST plays in the process. It is important that a CST develop an understanding of project administration. As with the other aspects of project management, a CST's level of experience and the structure of the PMT determine to what degree they are responsible for project administration.

6.0 DOCUMENT MANAGEMENT

In the project administration world, document management methods and practices regulate the development, approval, use, distribution, and retention of documents. Today, most PMTs use a combination of hard copy and computer-based applications in their document management practices. A CST is expected to develop an understanding of these methods and practices at a given location or company.

- Initiation documents: project scope, project management procedures, budget, PMT roles and responsibilities, communications plan, and project approval process
- Planning documents: schedules, and work orders
- Communications: reports, memos, emails

- Reference materials: prints, drawings, instrument lists, calibration sheets, vendor materials, and training materials.

A CST should understand how the Document Management System (DMS) is structured and used, which includes:

- Where the various types of documents are stored in the system
- Procedures for adding a new document to the system
- Procedures for checking out a document from the system
- Procedures for returning documents to the system

Given the wide variety of document formats and regulatory requirements, companies are turning to computer-based systems that track and store documents electronically. Often these systems allow users to store a document in the original format and in an open standard format. As an example (Figure 6-1), the system could contain a letter in Microsoft Office Word and a copy of the same letter in an open document format such as Adobe Portable Document Format (PDF).

FIGURE 6-1
Example of a File Stored in Proprietary and in Open Format

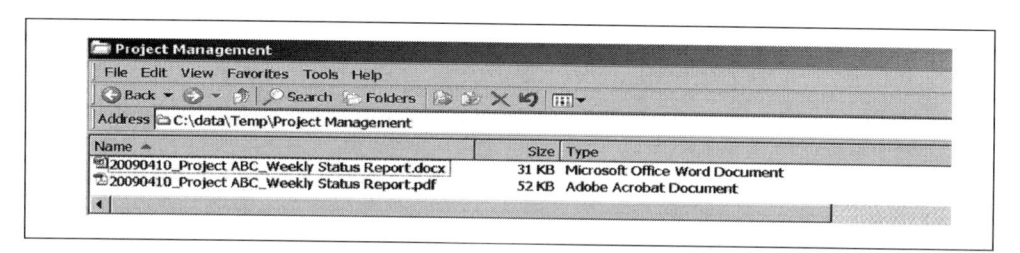

Name	Size	Type
20090410_Project ABC_Weekly Status Report.docx	31 KB	Microsoft Office Word Document
20090410_Project ABC_Weekly Status Report.pdf	52 KB	Adobe Acrobat Document

The proprietary format copy of the file can be updated or edited. Creating a second copy of the file in an open document format increases the likelihood of being able to later view the document. This is necessary because software manufacturers frequently change versions of their products, which, after several generations of updates, can introduce compatibility issues with older versions.

Project Administration

6.1 DOCUMENT ORGANIZATION

Regardless of their size, projects tend to generate large numbers of documents. The DMS brings order to the process, allowing the PMT to quickly store and retrieve documents. Again, it is important to note that the trend is moving away from hard copy documentation. That said, today the PMT must structure the document organization process to handle both paper and electronic files.

The first step is creating a document log, which allows the PMT to store and maintain information in an orderly fashion. As shown in Figure 6-2, a document log lists key information on each document. Created during the Initiation phase of the project, the document log design should incorporate any existing standards or practices of the customer. Strict adherence to the log format allows the PMT to maintain project documentation in an organized manner.

FIGURE 6-2
Example of a Document Log

Reference Number	Date	Author	File Format	File Name	Location	Comments
1.00	20090600	Document Log	Excel	20090701_HL_Unit 8 Control System Replacement Document Log	H:\HL\Unit 8 Control Project\Initiation Documents\	
2.00	20090701	Project Scope	Word	20090701_HL_Unit 8 Control System Replacement Project Scope	H:\HL\Unit 8 Control Project\Initiation Documents\	First Draft
2.01	20090815	Project Scope	Word	20090801_HL_Unit 8 Control System Replacement Project Scope	H:\HL\Unit 8 Control Project\Initiation Documents\	Approved final version
3.00	20090705	Print Index	Microstation	20090705_HL_Unit 8 Instrument List	H:\HL\Unit 8 Control Project\Reference	
4.00	20090814	Project Communications Plan	Word	20090814_HL_Unit 8 ControlCommunication Plan	H:\HL\Unit 8 Control Project\Communications Documents\	
5.00	20090820	Instrument Index	Excel	20090902_HL_Unit 8 Instrument List	H:\HL\Unit 8 Control Project\Reference\	
6.00	20091005	Vendor A RFP Response	Hard Copy	N/A	File Cabinet 103-Project Communication Documents	

A CST needs to understand how to use the document log. Over the course of a project, the PMT will generate many versions of some documents. Particular importance must be paid to the reference number to ensure that the current version of a document is used.

6.1.1 DOCUMENT REVIEW

In order to manage document review, the PMT must establish a formal document review process. Figure 6-3 highlights how the document review process consists of a series of steps, which are performed in sequential order. As defined by the PMT, authors create documents that are reviewed and ultimately approved by a specified individual or individuals.

FIGURE 6-3
  Example of a Document Review Process

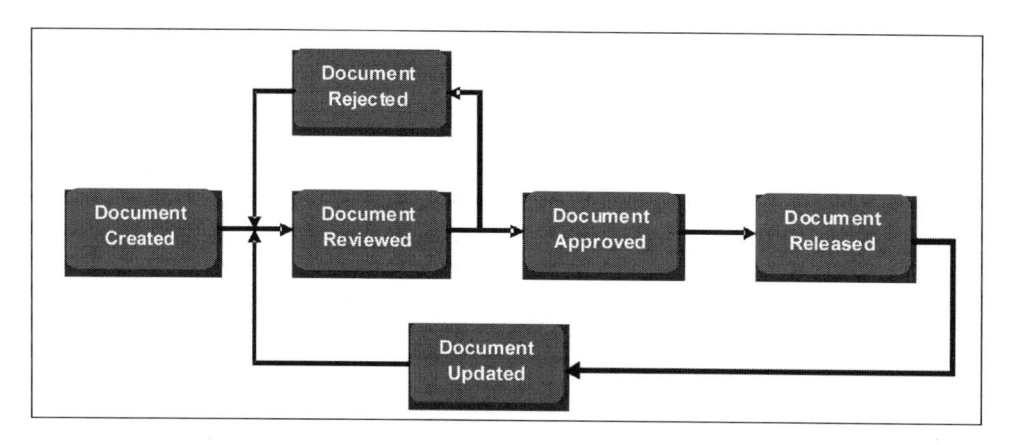

All project documents, regardless of type, must go through the same review process. For the process to work, the PMT must pay strict attention to the document management process and clearly communicate to all stakeholders.

Many project documents, such as prints, go through the process several times during the project lifecycle. This is due to design changes, material shortages, and change orders. Depending on the complexity of the change(s), one or more of the engineering disciplines may need to redesign, review, and approve the proposed change(s). Often a CST will collect field information, verify prints and drawings, and support the work of the engineers.

The document log plays a key role in the document approval process, as with each revision the PMT creates a new reference number for the document. Today many PMTs utilize web-based tools such as Microsoft SharePoint for document management. Products like this create an infrastructure that unites the various PMT disciplines, encourages collaboration, and simplifies the control aspect of document management.

6.1.2 DOCUMENT APPROVAL PROCESS

The document approval process defines who must approve a given document before it is released. While all documents go through the same process, the level of approval varies by the type of document and the organizational structure. As an example, for legal or regulatory reasons many documents require the approval of a licensed professional engineer.

Many companies have established policies for document approval for a facility, a business unit, or the corporation. The policy defines which Subject Matter Expert (SME) and stakeholders must approve a document. The individuals concerned could be internal or external to the facility where the project is being conducted, and may include engineering, technical, construction, maintenance or operating staff.

In all cases the PMT should clearly define the document approval process during the initiation phase of the project. A CST must develop a working knowledge of the document approval process for the project they are working on.

6.1.3 COMMUNICATIONS

As documents are created, reviewed, approved and released, the author is required to be the PMT and stakeholders. The Project Communication Plan described in Chapter 3 establishes guidelines for all project communications. While it is important to track all communications, it is even more so for items like prints and change orders that require approvals. Following the plan is the best way to ensure that everyone has input into the process and, more importantly, is working with the current version of a document. During the project a CST will have many communication opportunities as they complete equipment installations, calibrate instruments, update prints, and observe or sign off on work performed by contract staff. Therefore, a CST must understand how the document control communication process works and why it is important to project management.

6.2 EQUIPMENT & PERSONNEL RECORDS

Some documents, such as equipment and personnel records, may contain information that is proprietary and/or confidential. In the United States, laws such as the Data Protection Act of 1988 and the Health Information Privacy Act of 1996 make it illegal to share sensitive personal data or individually identifiable health information. In some situations, licensing agreements may require equipment operating manuals and software to be treated as confidential. To protect copies of these records, the PMT must put in place security and management procedures. These should provide the necessary protection for sensitive documents while still allowing access to the other records in the system.

As more and more companies turn to digital solutions for information management, the information technology world has developed systems to track and manage the use of all information. These systems authenticate the user's right to be on the system by using a password, a file on the user's computer, or a software hard key plugged into a USB port on the user's computer. In most cases authorized users can access the systems through the company's intranet or through the Internet.

6.3 PRINT MANAGEMENT

While the majority of prints are created during the planning phase of a project, they require maintenance until the conclusion of the project. Design mistakes, equipment delivery problems, and project change orders require the PMT to rework prints. As with the other project documents, prints are created or updated, reviewed, approved, and released following the established document management process for the project.

Following the design phase, the PMT obtains a master copy and working copy of all prints. Typically these are kept in separate locations so the PMT does not mistakenly alter a master print. In some cases, multiple working copies of a print are necessary due to the complexity of the project or the number of groups working in a particular area or system.

The creation and use of an electronically generated prints drawing has added another level of complexity to the print management

Project Administration

process. It is imperative that all copies be logged and tracked so that one group does not make a field change which would put equipment or personnel at risk.

While CSTs typically do not produce prints, they do play an important role in the print management process. As an example, a CST is provided a working copy of a print that contains information necessary to install a device. Once installation is complete, the CST should update the print and correct any errors. This process of updating and correcting prints is commonly referred to as "redlining" a print. On hardcopy (paper) prints, colored pencils are used to highlight the action. Typical colors used in the process are as follows:

- Additions: drawn in red
- Deletions: marked out using silver
- Comments to the drafter: printed in blue
- Changes circled in green and initiated by the person making the changes

Please note that white is never used in the redlining process.

At times a drawing may not have enough space to allow for redlining. If this is the case, redlines (changes or additions) may be drawn on a separate piece of paper and taped to the drawing. Figure 6-4 shows an electrical drawing that has been redlined.

FIGURE 6-4
Example of the Print Redlining Process

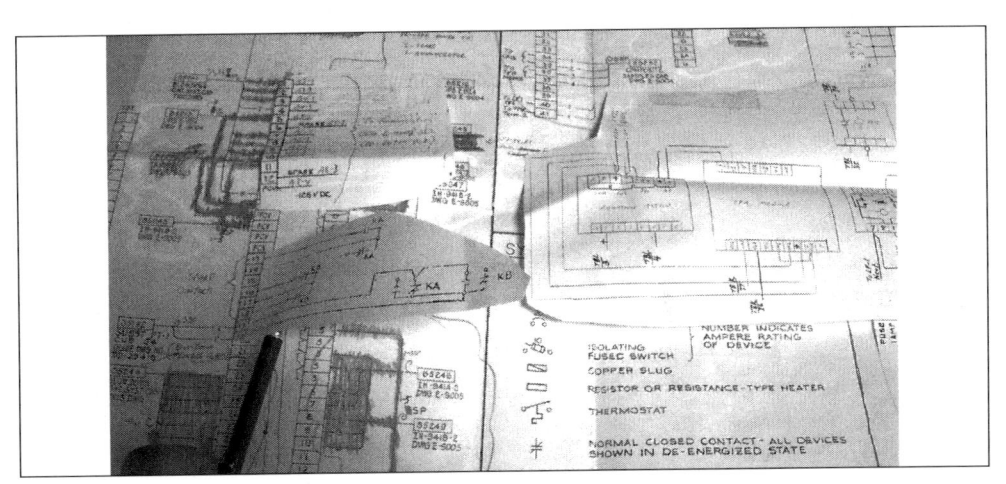

Today many CSTs are trained in drafting and the use of CAD (Computer-Aided Design) software. Some companies allow trained CSTs to redline using the CAD system. Note that the same rules as apply to paper, apply to redlining CAD drawings. When revisions are created and maintained electronically a redline file is created. Once the redline process is complete, a hard copy is plotted and the drawing is submitted for review. Following the review and approval process the correct drawing is released for use.

The final phase of print management is the creation of the "as-built" drawings. When the PMT has determined that work on a system is complete, the drawings are updated to reflect the changes that occurred since the design was finalized. This is a critical step for ensuring the safety of personnel and equipment. A CST may participate in the field review of "as-built" prints. In this process a CST verifies that the drawing represents the actual field wiring or equipment installation.

6.4 ARCHIVING PROJECT RECORDS

Records for a typical project consist of documentation in numerous types and formats. An organized system for archiving project records provides structure that makes the information usable. While PMTs continue to use hard copy archiving of project records, change is coming. As we have seen, more and more companies are migrating to electronic means of archiving project documents. Factors driving this change include:

- Reduced labor costs involved in processing files
- Reduced costs associated with maintaining physical copies of documents (space, labor, and security issues)
- Instant search and retrieval of documents
- Ability to view, print, fax, and email documents
- Back-up copies are located in a safe location, allowing quick restoration in the event of a system failure or natural disaster

Another driver is the fact that equipment and system manufacturers are now providing all manuals and other materials electronically.

Regardless of which storage system is used, the process is similar. Documents are placed in the system according to the type of document. As an example, a simple archiving system within the DMS would include sections for the following:

- Correspondence
- Prints & Drawings
- Vendor Materials

The actual archive structure may vary by company and type of project. A CST needs to become familiar with the system used on the project.

SUMMARY

Managing documents is one of the biggest challenges facing the PMT. Factors contributing to the problem include the diversity of project documentation, the ease with which a document can be created or changed, lost or misplaced documents, document version control, and the need to share documents with PMT members who often are not at the project site. Computer-based DMSs unite paper and digital documents in a format that allows the PMT to manage, control, search, distribute, and archive documents.

A CST needs to understand the principles of document management as they relate to each project. These include:

- What documents are required for the project
- How each type of document is created, reviewed, approved, and archived
- What is the distribution process for approved and revised documents
- What is the process to dispose of a document

Knowledge of these principles helps the CST to fulfill their duties on the PMT.

REVIEW QUESTIONS

1. Document management methodologies and practices do not regulate which of the following:
   A. Development Approval
   B. Software
   C. Use
   D. Distribution
   E. Retention

2. In project management the acronym DMS means:
   A. Data Memory System
   B. Data Management System
   C. Document Management System
   D. Desktop Management Suite

3. The Adobe Portable Document Format (PDF) is an example of an _____ file format.

4. While the trend is moving away from hard copies to electronic files, the PMT must structure the _____ to handle both types of files.

5. The _____ in the document log tells users what the current version of a document is.

6. Project documents such as prints go through the document review process one time during the project lifecycle. True/False

7. For legal reasons, access to _____ records must be limited.

8. The process of correcting or changing prints is commonly referred to as a print _____.

Project Administration

9. Which of the following colors is never used when redlining a print?

   A. Red
   B. White
   C. Blue
   D. Green

10. It is permissible to draw corrections or changes on a separate piece of paper and tape that over the portion of the drawing to be corrected. True/False

11. To improve document management and reduce costs, companies are migrating from hard copy archiving of project documents to _____ systems.

12. Which of the following would not be included in a simple archiving system?

    A. Correspondence
    B. Prints & Drawings
    C. Vendor Materials
    D. Company History

13. For the documents approval process to work, the PMT must clearly communicate to all _____.

7
PROJECT CLOSEOUT

As shown in Figure 7-1, Project Closeout is the last major phase of the project lifecycle.

FIGURE 7-1
Project Lifecycle

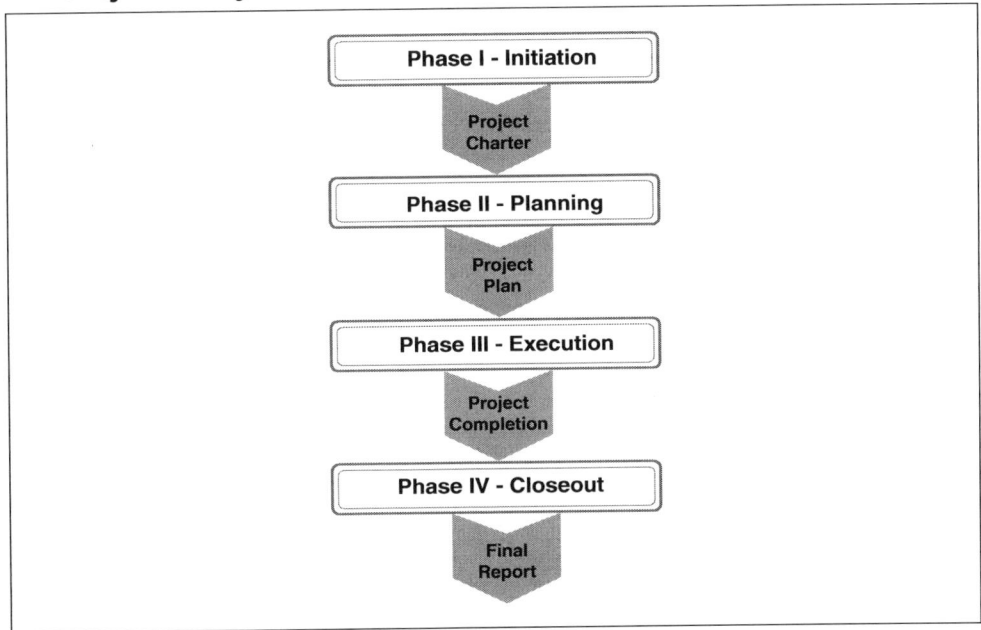

```
Phase I - Initiation
        ↓
   Project Charter
        ↓
Phase II - Planning
        ↓
    Project Plan
        ↓
Phase III - Execution
        ↓
  Project Completion
        ↓
Phase IV - Closeout
        ↓
    Final Report
```

During this phase, the PMT's work focuses on administrative tasks associated with bringing a project to closure. Key activities in this phase include:

- Obtaining owner/operator approval of the project
- Transferring responsibility for operations and maintenance to the owner/operator
- Transferring equipment and materials from the PMT to the owner/operator

- Transferring owner/operator staff members who were assigned to the PMT
- Conducting a "Lessons Learned" review of the project
- Completing and submitting the project closeout report
- Updating and archiving project records.

While the work of some members of the PMT is winding down, a CST will find much to do during the project closeout phase. This chapter examines the various activities of project closeout and the role of a CST.

7.0 TERMINATION

The first step in project closeout begins when the PMT and the owner/operator decides that the project is complete. Acceptance is based on criteria listed in the project plan and often in the contract. In many cases formal testing is required as proof the project performed in accordance with the project plan. This is known as "Customer Acceptance Testing." A CST may be involved in setting up devices or systems to collect customer acceptance testing data. In some cases, a senior CST may be responsible for conducting acceptance tests and preparing associated reports as required. This is a formal process which requires the PMT to obtain written acknowledgement from the appropriate level of owner/operator staff member(s).

No matter how successful the project, there are always issues that show up during the project termination process. Typically these are discovered as part of the formal acceptance testing or previously, as the owner/operator staff members test the functionality of the project. The PMT documents these issues in one or more "punch lists." An example of a punch list item might be a control valve that does not fully open. The PMT may send a CST to investigate and resolve the situation. Once the valve is functioning properly, the item is checked off the punch list.

Project Closeout

7.1 TRANSFER OF OWNERSHIP

Depending on the scope of the project, transferring ownership for the project can be a complicated process. It is easy to see that ownership includes responsibilities for the maintenance and operation of the project. However, in the case of a large project, these activities can include the transfer of materials and manpower. In most cases, the transfer of ownership is complete when the PMT leader (the Project Manager) receives written acknowledgement from the owner/operator stating that the project deliverables were met.

7.1.1 RESPONSIBILITIES

The first and most important stage of the transfer of ownership is owner/operator acceptance of the maintenance and operation responsibilities associated with the project. Working together, the PMT and the owner/operator develop procedures for the transfer. A CST serving on the PMT may support these efforts by conducting training for the owner/operator staff members and support other activities designed to facilitate the transfer of knowledge from the PMT members. Examples range from helping train the facility operations personnel on the use of the control system and to teaching the owner/operator technicians how to service and maintain newly installed analyzers.

7.1.2 DOCUMENTS

As discussed in Chapter 2, project documentation covers a wide variety of items. The PMT collects or archives these items to create a historical record of the project. Until recently, all project documentation was archived in hard copy form. Today, as we have seen, companies are turning to electronic archiving in a move to reduce costs, improve accessibility, and increase accuracy of the documentation. The software package IXOS is another example of a document management platform. This system works with the enterprise business suite such as SAP. Systems such as IXOS allow users to archive documents in a proprietary format and in an open source file format. Figure 6-1 (Chapter 6) shows an example of a weekly project status report archived in the proprietary format (Microsoft Word .docx format) and in the open source format (Adobe Acrobat Document .pdf).

CSTs are often called upon to review project documents prior to transfer to the project archives. This work is extremely valuable, as the PMT tries to ensure that the documentation correctly represents the project in its completed form ("as-built"). Prints, calibration records, and spare parts inventory lists are a few examples of the types of documents a CST may review.

7.1.3 EQUIPMENT AND MATERIALS

The amount of effort required for the transfer of equipment and materials is directly related to the complexity of the project and the requirements of the project plan. During the project the PMT may acquire spare parts, specialized materials, and equipment that will be transferred to the owner/operator at the end of the project. A project to upgrade a power plant control system from pneumatics to a Distributed Control System (DCS) is an example of a complex project. This type of project often results in a large amount of materials being transferred from the PMT to the owner/operator, because a DCS requires much different technology to maintain and operate than a pneumatic system.

A CST will often support efforts to inventory and transfer technical materials to the owner/operator. A CST may inventory spare parts for the control system and associated field devices and equipment. Once the inventory is complete the CST may work with the agents of the owner/operator who have the authority to sign off on the transfer of materials. As with other tasks, the scope of this work should be identified in the project plan.

7.1.4 FACILITIES

Facilities used by a PMT during a project usually fall into the following categories:

- Owned by contractors
- Existing facilities owned by the owner/operator
- Newly constructed facilities that are part of the project.

Project Closeout

Most of the facilities owned by contractors (for example, site offices) are mobile and are removed once the project is complete. A CST may be involved in working with the owner/operator staff to remove utilities (electricity, water, sewer, and land telephone lines) from these facilities. The PMT must also ensure that existing facilities are returned to the condition they were in when first occupied. Documents such as a contractor's handbook or similar facilities guidance documentation typically cover the expectations of the owner/operator. In a similar manner, the project plan provides the PMT guidance for newly constructed facilities.

7.1.5 PERSONNEL

Some companies contract for turnkey projects. The makeup of the PMT on this type of project varies from limited participation of the owner/operator staff to a PMT that is made up entirely of external members. However, many companies utilize a PMT that may include members from other facilities, members from corporate headquarters, and members of the facility owner/operator staff. The project plan should include the process for releasing the PMT members once their work on the project is complete. This is especially important for PMT members that are owner/operator's employees, as they usually have other duties at the facility.

Depending on the length of the project, the PMT may be asked to provide information on PMT member performance. If this is the case, the PMT must record performance-related information. This is especially true when a PMT member's performance is not acceptable. As an example, a group of owner/operator technicians are assigned to the PMT. A senior CST may be assigned to supervise these individuals while they are conducting the Factory Acceptance Test (FAT) at the system provider's (manufacturer or control system integrator) location and while performing other project-related activities at the owner/operator's facility. During this procedure, the senior CST should record any performance-related issues. PMT member performance evaluation should also include recognition for individuals whose efforts are above expectations.

The last step in releasing personnel is ensuring that the individuals turn in equipment and materials that are part of the project. The best way to handle this step is to inventory the member's personal equipment at the start of the project, and then inventory it again at the end. This might sound like a lot of trouble, but it can prevent an individual from accidentally leaving the project site with an essential piece of equipment. Examples include hard keys to software, specialized cables, and calibration equipment. Software should not be overlooked during the pre- and post-project inventory.

7.2 LESSONS LEARNED SESSION

All projects have areas of success and areas that could be improved upon. The process of reviewing and documenting these areas is often referred to as a "Lessons Learned" session. All members of the PMT and members of the owner/operator staff who were associated with the project should participate in the Lessons Learned session.

Some industries conduct an "after-action review," which is the process used by the U.S. Military. No matter what you call the process, the intent is to document the following for each major project activity:

- What took place
- Reasons for the particular outcome
- Steps that would have improved the outcome

Many companies share the results of the Lessons Learned session throughout the organization. Not only can this improve similar projects at other facilities, but knowing the strengths and weaknesses of the PMT members and owner/operator employees who were associated with the project helps in planning similar projects.

During a typical automation project, a CST often begins during the initiation phase of the project and can continue through the early stages of closeout. Each activity presents the CST with the learning opportunities. As an example, a CST may gain valuable experience installing, calibrating and testing a data acquisition system. Capturing this direct knowledge of the project equipment, systems, and the owner/operators facilities is important to the Lessons Learned process.

7.3 PROJECT CLOSEOUT REPORT

The final activity for the PMT in completing the project is the closeout report. The report serves as a historic record of:

- Project closeout tasks completed
- Performance of the PMT
- Overall project performance
- Project conformance to the project plan
- Lessons Learned
- Recognition for PMT members

The task of preparing the project closeout report typically falls to the Project Manager, who gets input from the PMT, the owner/operator, and any other stakeholders. Other than providing input to the Project Manager, a CST rarely will have much to do with the actual creation of the closeout report.

SUMMARY

In recent years, industry has elevated the importance of the project closeout phase in an effort to continually improve performance. By design, each project's closeout phase explores key aspects of the project, which can reveal opportunities for improvement. The work also highlights positive aspects of the project that should be repeated on future projects. The entire process ends with the submission of the project closeout report, which documents these findings.

The project closeout phase is the final push for the PMT and as such can be very stressful. It also presents many learning opportunities as the team works to solve problems and transfer responsibilities to the owner/operator staff. The CST should clearly understand their role in this phase of the project and look for opportunities to learn.

REVIEW QUESTIONS

1. The Closeout Phase focuses mainly on _____ tasks.

2. The PMT may conduct a _____ test to prove that the project can perform in accordance with the project plan.

3. System performance problems discovered during closeout are recorded on a _____ _____.

4. Which of the following items is not typically transferred during the closeout phase?
   A. Ownership (operations & maintenance)
   B. Facilities
   C. Project management responsibilities
   D. Owner/operator personnel assigned to the project
   E. Project documentation
   F. Goods and materials

5. Transfer of ownership is complete when the Project Manager receives _____ confirmation that the project is complete.

6. On large projects transfer activities can include the transfer of materials, documentation, and _____.

7. Which item is not a reason companies are turning to electronic archiving of project documents?
   A. To reduce costs
   B. To improve accessibility
   C. To improve accuracy
   D. To eliminate field verification of as-built drawings

8. A CST conducting training for the owner/operator staff members is an example of _____ transfer.

Project Closeout

9. At the end of the project, the PMT typically _____ spare parts, specialized materials, and equipment to the owner/operator.

10. The _____ _____ details the process for releasing the PMT members once their work on the project is complete.

11. The expectation of the owner/operator as to the condition of _____ (new and existing) at the termination of the project should be defined in the project plan.

12. The Lessons Learned process attempts to document all but one of the following items:
    A. What took place
    B. Reasons for the particular outcome
    C. Estimated costs to improve performance
    D. Steps that would have improved the outcome

13. The final activity of Phase IV is the _____ _____, which serves as a historic record of all aspects of the project.

8

AN AUTOMATION PROJECT

The first seven chapters of this book covered the project organization and administration areas listed in Domains VI and VII of the ISA CCST test. This information should prove valuable to candidates preparing for the CCST test. This chapter provides an example, based on an actual project, of the duties and responsibilities that might be assigned to two technicians who are members of a PMT. The project is the replacement of the analog electronic control system at a fossil fuel fired power plant with a distributed control system (DCS).

Our first CCST, Jim Edwards, works as an instrument and controls technician at the power plant. He is a level II CCST who has seven years of experience. Our second CCST, Dan Devine, is a senior technical specialist who works in the corporate engineering department of the company who owns the power plant. He is a level III CCST with thirteen years of experience.

Readers should keep in mind that actual project assignments for CCSTs vary by industry, the availability of resources, and the nature of the project.

8.0 PHASE I – INITIATION

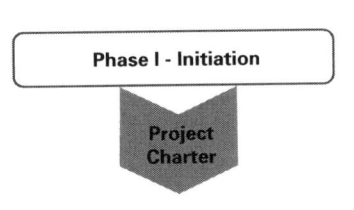

During the initiation phase of a project, the majority of the activities performed by a CCST involve gathering information. While some corporate-related materials are used in the process, the majority of the information is in site-specific documentation. Examples of this documentation include prints, vendor manuals, and procedures specific to plant operations, safety, and security. As Technical Lead on the project, Dan works with members of the plant staff to obtain this information. This information is used to build a business case for the project.

Once the project is approved the Project Manager establishes a PMT. For this project, the PMT includes members from the corporate engineering department, the power plant staff (operations and maintenance), the selected A&E (Architecture and Engineering) firm, and the control system manufacturer. Jim is one of the members, representing the power plant maintenance department on the PMT. Typically Jim and other CSTs assigned to the project will serve on an as-needed basis until the execution phase of the project. During the initiation phase of the project, their primary role is to assist the PMT in obtaining site-specific documentation.

8.1 DESIGN

The analog electronic control system was installed in late 1968. Since then the plant has replaced many outdated field devices and modified some systems. While the design efforts are led by the A&E firm, the process requires a good deal of manpower from the PMT and the power plant staff. The PMT members review the existing system and, using this information, develop a design. This information is used to produce the Request for Proposal (RFP) and subsequent Request for Quote (RFQ).

As Technical Lead on the project, Dan works closely with the power plant staff to ensure that accurate information is included in the design process. During this phase of the project, Jim can expect to receive numerous questions on the existing control system. In many cases, Jim and the other technicians assigned to support the project will conduct field verification in order to ensure that the information on the existing system is correct.

8.2 PHASE II – PLANNING

The project planning phase begins once the PMT is in place and the design phase is complete. In the planning phase, the PMT develops the project plan that serves as a guide for all project activities. The project plan will include:

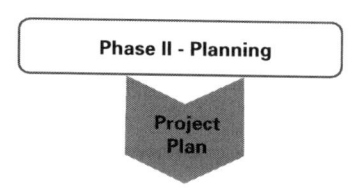

Phase II - Planning

↓

Project Plan

An Automation Project

- Resource allocation (labor, materials, and equipment)
- Financial matters (budget and reporting requirements)
- Quality assurance (monitoring and control processes and reporting)
- Document management
- Communications
- Project closeout

In most cases, Jim and the other CSTs assigned to the project will not have much input into the planning phase of the project. On the other hand, as Technical Lead, Dan is a member of the PMT and will be involved in planning activities in the areas of resource allocation, quality assurance, and project closeout.

8.2.1 VENDOR SELECTION

Today most companies have established supply chain guidelines for the sourcing of contract services, materials, and equipment. Typically this involves submitting a Request for Proposal (RFP), evaluating the responses, submitting a Request for Quote (RFQ), and selecting a vendor. The PMT works closely with the Purchasing Department to develop and perform technical reviews on RFQ documents. During vendor selection, Jim may be called upon to provide information required to answer a vendor's questions (see Figure 8-1).

FIGURE 8-1
Sample Vendor RFQ Question

> The RFQ documentation indicates that the existing boiler temperature measurement sensors are 10 ohm copper RTD's. Are these two- or three-wire RTD's? Given the age of these devices, please verify that they have not been replaced with another type of temperature sensor. If they have been replaced, please provide the engineering details on the replacement devices.

As the Technical Lead, Dan serves on the RFQ Technical Review Team for automation vendors. The team works to ensure that vendor selection is based on technical merit and best value, not lowest price.

8.3 PHASE III – EXECUTION

Until this phase, a CCST typically does not work full time on the project. Once the project enters the execution phase, most of the PMT members are dedicated full time to the project. In fact, it is in this phase that PMT members are most likely to work overtime in support of factory acceptance testing, installation, start-up, and customer acceptance testing.

8.3.1 FACTORY ACCEPTANCE TESTING

As the name implies the Factory Acceptance Test (FAT) takes place at the manufacturer's or system integrator company's assembly facility. A well-planned and executed FAT is an essential element of a successful project. For this project, the FAT team members include:

- A Project Manager from the owner's corporate engineering department
- A Senior Technical Specialist from the owner's corporate engineering department
- Two automation technicians, two electricians, and one operator who are employed by the owner and work at the project facility

During a typical FAT, the PMT members will simulate all inputs to the control systems. This point-to-point check quickly identifies wiring and programming problems. The team will also review the control system graphics to ensure that the screen design and overall system layout meet the needs of the plant operation and maintenance departments. Since the test is conducted at the assembly facility, the technicians who assembled the system can quickly work to resolve any issues. While the FAT is a time-consuming process, it is far less expensive to resolve problems at the assembly facility than it is at the power plant.

As the Technical Lead, Dan will supervise the FAT team members who are actually conducting the tests. His role is to ensure that the team follows the FAT plan and documents their work. This includes marking up the various control system prints and providing the management staff at the assembly facility with a list of items that do not perform to design.

Working with other members of the FAT team, Jim will simulate inputs to the control system and document that the system responds correctly. At times, Jim and the other FAT team members may assist the employees at the assembly facility to troubleshoot wiring and programming problems with the control system. This is an excellent opportunity for Jim and the owner/operator staff members who are assigned to the FAT team to learn about the control system.

8.3.2 INSTALLATION

For this project the SIC and subcontractors will install the control systems and related equipment. During this phase, Dan will monitor the work to see that it meets the project plan.

Depending on the workload at the power plant, Jim's project activities may be limited during the early stages of installation. In some cases, Jim and the other CSTs assigned to the PMT may be called upon to shadow (oversee) the subcontractors. Once the installation is complete, Jim, the other CSTs, the electricians, and the operator on the PMT will perform loop calibration on all field devices. This ensures that all points in the control system are receiving signals from the correct device. This activity also provides the PMT documentation on the calibrations of all devices including the critical trips and interlocks. This experience will serve Jim well once maintenance responsibilities are turned over to the plant.

8.3.3 START-UP

Even with a well-thought-out plan, start-up tends to have an element of organized chaos. It is the role of the PMT to bring order to this process to ensure the safety of the personnel and equipment. A project engineer, technical lead, or possibly the project manager works with the Operations Department to thoroughly test each system before being released to service. Once system testing is complete the generation unit start-up begins.

For this project, start-up comes in stages as testing determines that major components, such as the boiler and turbine, are ready for operation. With each stage the controls and systems are tuned and adjusted to allow proper operation.

The final stage of start-up consists of a performance test, which verifies the control system ability to performance to conditions specified in the contract. The requirements of this test are defined in the project plan and included in the control system contract. As an example, the contract may include a requirement for the control system to be able to raise load on the unit by ten megawatts per minute. The test consists of sending a demand signal to the control system, monitoring the major systems, and documenting that the control system did achieve the objective of maintaining stable operation at the desired rate of load change.

As in the installation stage, start-up presents a host of learning opportunities for a CCST. Start-up testing is a somewhat unpredictable process that tends to require a good deal of overtime to keep the project on schedule. During this stage, Dan's activities include:

- Coordinating the work of the various groups with the Operating Department
- Ensuring that all changes are documented (prints, calibration records, equipment lists, etc.)
- Providing progress reports to the Project Manager

Jim works closely with the system integrator and other contractors during the start-up testing process. Having firsthand knowledge of the power plant makes Jim an excellent resource to the personnel working on start-up. Participation in start-up also presents Jim with an opportunity to support Dan in ensuring that changes to the systems are documented.

8.4 PHASE IV – CLOSEOUT

As presented in Chapter 7, project closeout consists of several key activities. The overriding objective of the closeout phase is to obtain acceptance

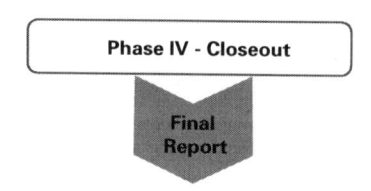

Phase IV - Closeout

Final Report

An Automation Project

of the project and to transfer operations and maintenance responsibilities.

Once the start-up activities are successfully completed the PMT seeks approval from the owner/operator for the project. As defined in the contract and listed in the project plan the owner/operator will accept the control system once it has successfully completed a series of performance-based acceptance tests. The testing is performed to standards such as the American Society of Mechanical Engineers (ASME) performance test codes. In the case of this project, the tests are conducted by an independent testing contractor who schedules the work with the owner/operator operations staff. The roles of the PMT members vary from actual support to observing the testing. Jim is assigned to support the testing contractor by providing information on the location of test connections on the unit. Dan is responsible to see that the equipment setup and calibration meet the testing standards.

Once the tests are completed, the PMT will evaluate the test data and if successful, present the test results to the owner/operator member who is authorized to officially accept the project. From this point forward, the plant maintenance and operations departments are responsible for the new control system.

Now that the project work has been accepted, the PMT moves on to transferring equipment, materials, and manpower to the power plant. Under Dan's direction, the PMT members inventory all equipment and materials related to the project. Examples of equipment include spare transmitters, thermocouples, control system modules, and calibration equipment. Examples of materials include consumable items such as replacement parts, calibration gases or fluids, wiring, and tubing. As the items are turned over, a goods receipt is signed by a responsible plant employee. This includes all project documentation. Since Jim works at the plant he knows whom to contact for the various items. As an example, he takes all the project documentation to the plant Business Support Supervisor, who is responsible for maintaining the Reference Room at the plant.

Just prior to releasing the PMT members (corporate, power plant, consultants, contractors), the Project Manager conducts a Lessons Learned session. Both Dan and Jim participate in this exercise. While the session is designed to capture the strengths and weaknesses of the project, it also serves as a learning experience for a CST as some issues are related to the knowledge level of the PMT. During the project, Jim installed, programmed, calibrated, and tested a DAS that supplies information to the control system. At the Lessons Learned session, he shares a problem he encountered when the DAS did not fit in the area shown on the installation print. He also shares a programming tip not mentioned in the manufacturers programming manual.

The final duty of the Project Manager is producing the project closeout report. Although the other PMT members are now gone, Dan is available if the Project Manager requests his support. The support he would provide would typically be limited to conducting research or finding documents to include in the report.

SUMMARY

This chapter presented readers with examples of the duties a CCST may perform during the four phases of a project. Serving on a PMT is an excellent learning experience for a CCST. In this project, both Dan and Jim gained experience that will serve them well in the future. As Dan moves on to the next control system project, he takes with him lessons learned that will help ensure the success of the next project. At the same time Jim has gained valuable knowledge of the design and operation of the new control system. His work on the PMT has introduced him to project management and the role a CCST plays in each phase of a project.

FIGURES

Chapter 2

EMERGENCY RESPONSE MANUAL

EMERGENCY PHONE NUMBERS — September 2009

Police: _____
Fire: _____
Medical: _____
Security Control Center: XXX-XXX-XXXX
Local Security Guard: XXX-XXX-XXXX

- INTRODUCTION/RESPONSE TEAM
- DUTIES & RESPONSIBILITIES
- TORNADO/CIVIL DISTURBANCE
- MEDICAL EMERGENCIES/CHEMICAL RELEASE
- FIRE
- MEDIA RELATIONS GUIDELINE
- BOMB THREATS
- WORK PLACE VIOLENCE
- TELEPHONE THREAT CHECKLIST
- BIOLOGICAL AGENT EXPOSURE
- INFORMATION TECHNOLOGIES/ENABLERS AT RISK
- ROUTINE PHYSICAL SECURITY GUIDELINES
- EVACUATION ASSEMBLE AREAS

Figure 2-2 Typical Site ERP Manual

Chapter 3

- Interpersonal Skills
- Writing Skills
- Oral Skills
- Facilitation Skills

Figure 3-1 Communication Skills

Chapter 4

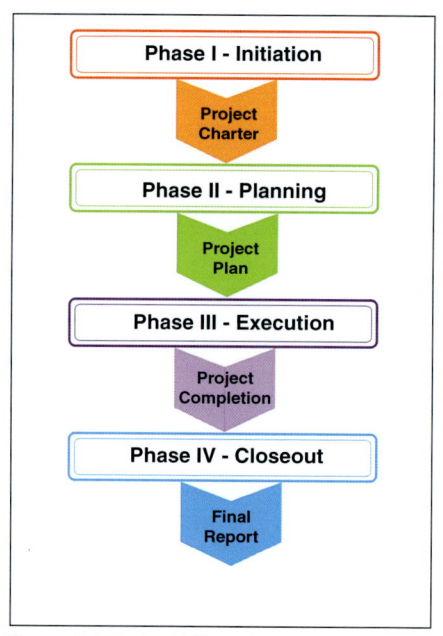

Figure 4-1 Project Lifecycle

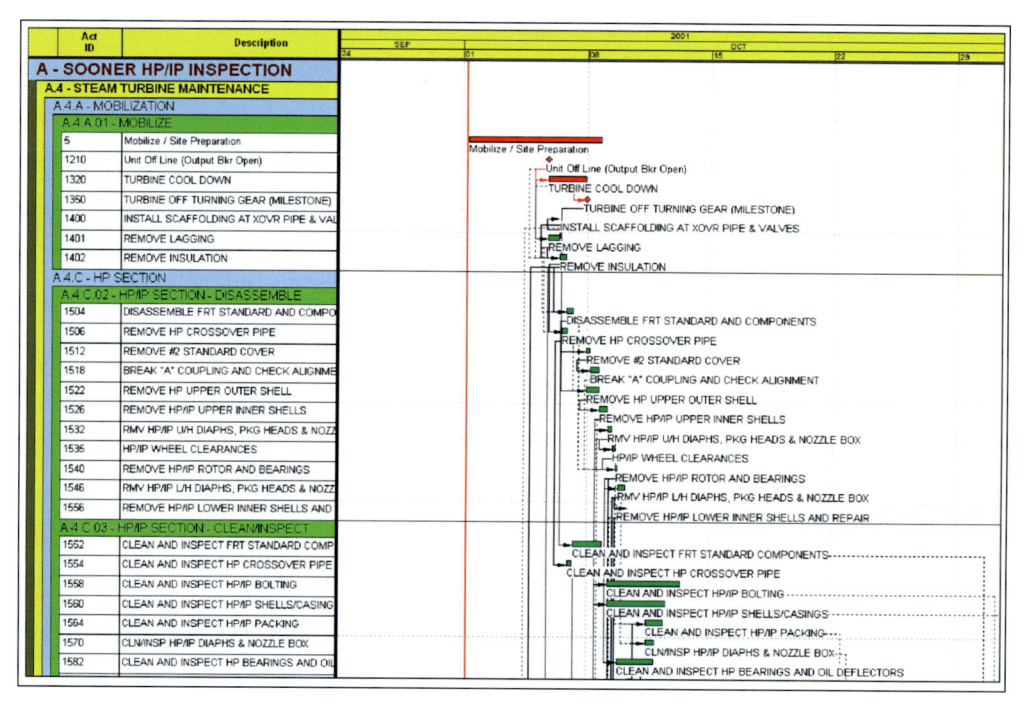

Figure 4-2 Gantt Chart

Figures

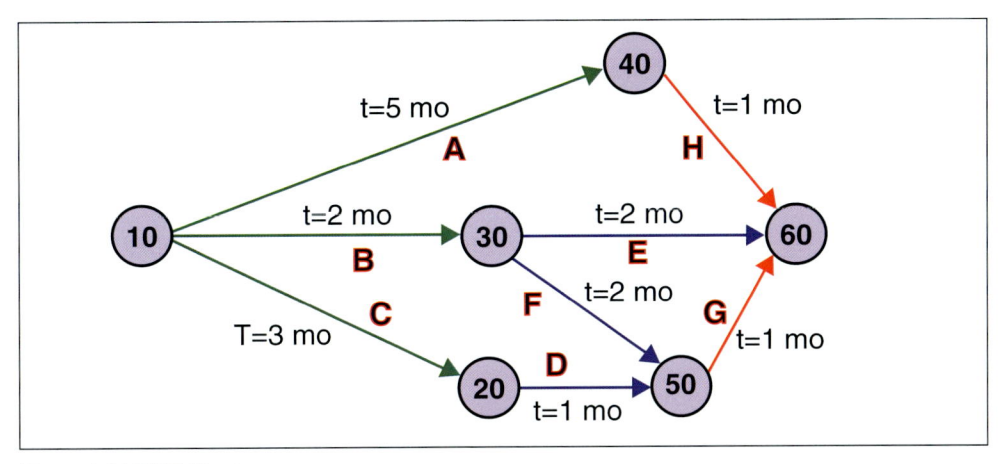

Figure 4-3 PERT Chart

Chapter 6

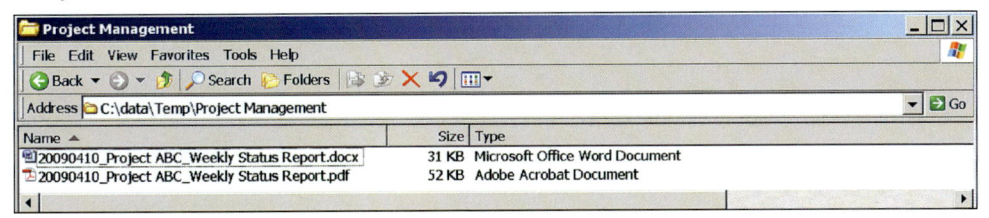

Figure 6-1 Example of a File Stored in Proprietary and in Open Format

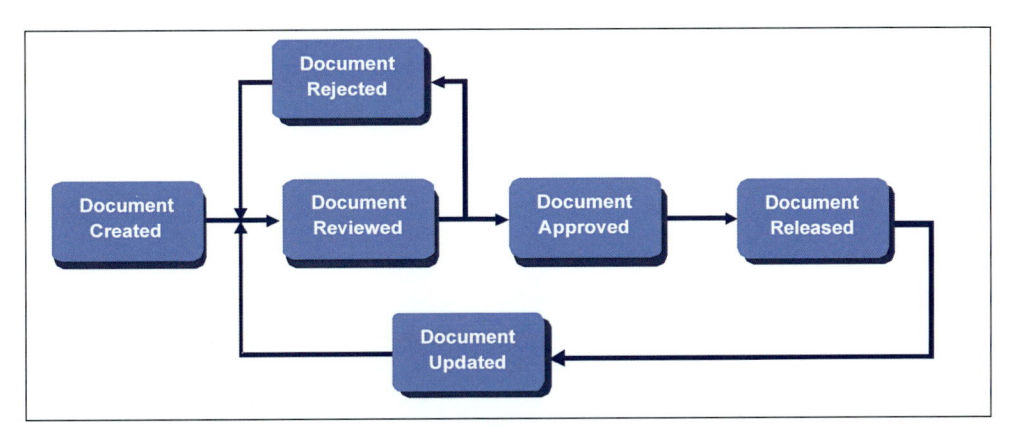

Figure 6-3 Example of a Document Review Process

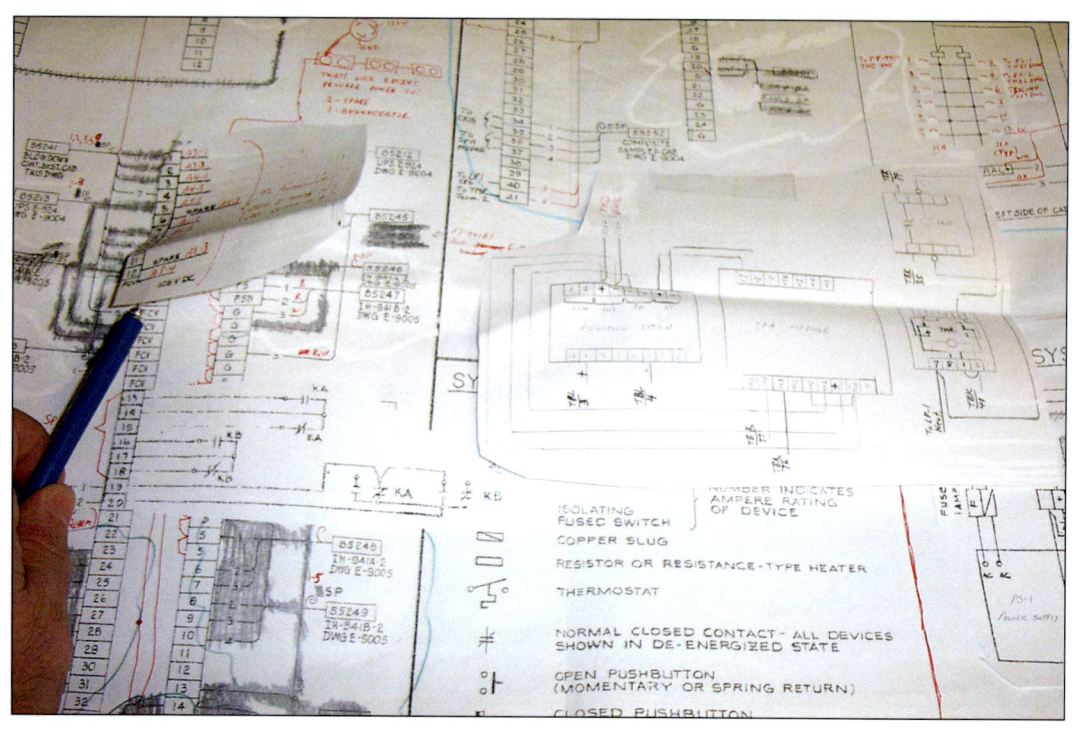

Figure 6-4 Example of the Print Redlining Process

ACRONYMS

A&E	Architects & Engineering Firms
CAD	Computer-Aided Design
CSE	Control System Engineer
CCST	Certified Control Systems Technician
CST	Control System Technician
DCS	Distributed Control System
DEQ	State Department of Environmental Quality
DHA	Department of Homeland Security
DMS	Document Management System
DOT	Department of Transportation
EPA	Environmental Protection Agency
ERP	Emergency Response Plan
ISA	International Society of Automation
MSDS	Material Safety Data Sheets
OSHA	Occupational Safety & Health Administration
PFD	Process Flow Diagram
PI&D	Piping and Instrumentation Drawings
PM	Project Manager
PMT	Project Management Team
PPE	Personal Protective Equipment
SIC	System Integration Companies
SMT	Subject Matter Expert
WAP	Work Authorization Permit

ANSWERS

CHAPTER 2

1. How can a CST obtain a copy of the Site Procedure for a location?

   Contact the site Operating Authority

2. Explain the purpose of, and information provided by, Material Safety Data Sheets (MSDS).

   The eight sections of the MSDS convey the hazards associated with a particular chemical to employees who work in or around that chemical.

3. What is the purpose and importance of lock-out/tag-out procedures?

   To limit the risk to personnel and equipment from the unexpected start-up of equipment or the release of stored energy.

4. What types of materials make up project engineering documentation?

   Various types of prints, drawings, lists, indexes, and vendor information

5. What is the purpose of an "as-built" drawing?

   Provide a record of as-built conditions as a reference for future maintenance and facility change processes.

6. A maintenance department typically maintains a _____ of prints for routine use.

    working copy

7. Logic Diagram depict _____ relationships between instruments and the process.

    logic symbols

8. Installation details display the power and signal connections between devices, but these do not correspond to the _____ of the devices.

    physical locations

9. How can physical security issues affect a project?

    Limit access to the facility and/or areas within the facility.

10. A _____ _____ system that tracks and stores electronic and/or paper documents.

    document management

11. Today many _____ _____ _____ include project management software?

    Enterprise software suites

CHAPTER 3

1. What does a project communications plan define?

    Who is responsible for the particular communication; why, when, how often, and in what format it is required.

Answers

2. What are the basic communications skills that a CST must master?

   The four areas of basic communications skills are interpersonal, written, oral, and facilitation.

3. Given that more and more project documents are produced electronically; does a CST really need good written communications skills? (Explain)

   Yes. The technology used to produce documents does not reduce the need for a CST to develop good writing skills.

4. When is it extremely important to have mastered good interpersonal communication skills?

   When situations become adversarial.

5. Should employees expect privacy for the communication materials (memos, text messages, email, etc.) they produce on the job?

   No, all materials produced on the job are the property of the company.

6. Why are oral communication skills important to a CST?

   Because a CST must be able to effectively present and defend their ideas.

7. Why does a new CST need to be familiar with the facilitation process?

   Knowledge of the facilitation process allows the CST to effectively lead a meeting.

8. Understanding the social attitudes and behaviors of the people working on a project allows a CST to avoid _____ _____ communication problems.

   culture-related

9. Why is it important for PMT members to identify an acronym or abbreviation by its full name?

   An acronym or abbreviation can have different meanings to the various disciplines.

CHAPTER 4

1. What information can a CST obtain from a project Gantt chart?

   The projected project tasks start and finish dates and the dependency relationships between the various tasks.

2. Today many companies are migrating from stand-alone project management software to _____ _____ suites.

   enterprise software

3. Project Management Team members can include:
   A. Customer (local site and corporate) employees
   B. A&E firms
   C. Construction companies
   D. Vendors
   E. Manufacturers
   F. All of the above

   F

4. A Project plan is developed during Phase _____.

   II

Answers

5. _____ involves the use of schedules to plan and manage a project and produce project status reports.

    Planning

6. A _____ chart presents a project schedule, in a manner that highlights the dependency of one task upon another.

    Gantt

7. One advantage of enterprise suite project management applications is that the application is _____ with the other enterprise applications.

    integrated

8. The project plan consists of a list of interdependent _____.

    tasks

9. As a member of a Project Management Team, a _____ is expected to understand how tasks are assigned and completed work is reported.

    CST

10. For a CST, the important time management issues include:
    - A. Location of the work
    - B. Equipment requirements
    - C. Support required to complete the task
    - D. Safety/security concerns that could impact the work
    - E. All of the above

    E

CHAPTER 5

1. What are some of the ways a CST contributes to the monitoring process?

   Completing work orders, verifying receipt of automation materials, and collecting status report information.

2. _____ software automates portions of the project monitoring and control process.

   Project management

3. On projects that utilize enterprise software suites for planning and scheduling tasks, a CST may log into the system and confirm a _____ once a task is complete.

   work order

4. Monitoring helps the PMT ensure that the appropriate _____ are available to support activities and that work is progressing according to the project plan.

   resources

5. _____ activities involve actions that ensure the project schedule, budget, and resource allocation align with the project plan.

   Control

6. When a CST validates shipping documentation against the equipment delivered, they are helping control the project _____.

   budget

Answers

7. Which of the following items are included in a project status report:
   A. Schedule
   B. Budget
   C. Risk
   D. All of the above

   All of the above

8. The actual methodology and reporting interval for the project status report is defined in the _____ _____ plan.

   project communications

9. Material delivery delays and subsystem incompatibility are potential _____ to a project.

   risks

10. The duties of a senior CST serving as Technical Lead on a project may include providing input to help in identifying risks and developing a _____ plan.

    risk mitigation

CHAPTER 6

1. Document management methodologies and practices do not regulate which of the following:
   A. Development Approval
   B. Software
   C. Use
   D. Distribution
   E. Retention

   B

2. In project management the acronym DMS means:
   A. Data Memory System
   B. Data Management System
   C. Document Management System
   D. Desktop Management Suite

   C

3. The Adobe Portable Document Format (PDF) is an example of an _____ file format.

   open

4. While the trend is moving away from hard copies to electronic files, the PMT must structure the _____ to handle both types of files.

   DMS

5. The _____ in the document log tells users what the current version of a document is.

   reference number

6. Project documents such as prints go through the document review process one time during the project lifecycle. True/False

   False

7. For legal reasons, access to _____ records must be limited.

   personnel

8. The process of correcting or changing prints is commonly referred to as a print _____.

   redlining

Answers 93

9. Which of the following colors is never used when redlining a print?
   A. Red
   B. White
   C. Blue
   D. Green

   B

10. It is permissible to draw corrections or changes on a separate piece of paper and tape that over the portion of the drawing to be corrected. True/False

    True

11. To improve document management and reduce costs, companies are migrating from hard copy archiving of project documents to _____ systems.

    Electronic

12. Which of the following would not be included in a simple archiving system?
    A. Correspondence
    B. Prints & Drawings
    C. Vendor Materials
    D. Company History

    Company History

13. For the documents approval process to work, the PMT must clearly communicate to all _____.

    stakeholders

CHAPTER 7

1. The Closeout Phase focuses mainly on _____ tasks.

   administrative

2. The PMT may conduct a _____ test to prove that the project can perform in accordance with the project plan.

   acceptance

3. System performance problems discovered during closeout are recorded on a _____ _____.

   punch list

4. Which of the following items is not typically transferred during the closeout phase?
   A. Ownership (operations & maintenance)
   B. Facilities
   C. Project management responsibilities
   D. Owner/operator personnel assigned to the project
   E. Project documentation
   F. Goods and materials

   C

5. Transfer of ownership is complete when the Project Manager receives _____ confirmation that the project is complete.

   written

6. On large projects transfer activities can include the transfer of materials, documentation, and _____.

   personnel

Answers

7. Which item is not a reason companies are turning to electronic archiving of project documents?
   A. To reduce costs
   B. To improve accessibility
   C. To improve accuracy
   D. To eliminate field verification of as-built drawings

   D

8. A CST conducting training for the owner/operator staff members is an example of _____ transfer.

   knowledge

9. At the end of the project, the PMT typically _____ spare parts, specialized materials, and equipment to the owner/operator.

   transfers

10. The _____ _____ details the process for releasing the PMT members once their work on the project is complete.

    project plan

11. The expectation of the owner/operator as to the condition of _____ (new and existing) at the termination of the project should be defined in the project plan.

    facilities

12. The Lessons Learned process attempts to document all but one of the following items:
    A. What took place
    B. Reasons for the particular outcome
    C. Estimated costs to improve performance
    D. Steps that would have improved the outcome

    C

13. The final activity of Phase IV is the _____ _____, which serves as a historic record of all aspects of the project.

closeout report

INDEX

Adobe Portable Document Format (PDF) 50
After-Action Review 66
An Automation Project 71
ANSI Standards Z400.1-2003 9
Architectural & Engineering (A&E) 34
Archiving Project Records 56
As-Built Drawings 56
Automation Projects 1
Automation Projects Engineering Documents 13

Basic Communications 21
Basic Duties of Other Disciplines 34
Budget 46
Business Software Suites 14

CCST Exam 2
Certified Control Systems Technician 2
Change Management 45
Change Order 46
Classification of Prints & Drawings 12
Clearance procedure 8
Communication 53
Communication Requirements 19
Communication Skills 21
Communication Techniques 26
Communications Plan 19, 25
Computer-Aided Design (CAD) 15, 56

Construction Companies 34
Control of Hazardous Energy 8
Controls 44
Corporate Politics 28
Cultural 30

Data Protection Act of 1988 54
Design 72
Diaries or logs 43
Document Approval Process 53
Document Log 51
Document Management 49
Document Management System (DMS) 15
Document Organization 51
Document Review 52
Document Review Process 52
Documentation Location and Media Types 14
Documents 63
Drawing Index 12

Emergency Response Plan 7
Enterprise Software Suites 14
Equipment & Personnel Records 54
Equipment and Materials 64

Facilitation Skills 25
Facilities 64
Factory Acceptance Testing 74
FAT Team 74

Gantt Chart 36
Ground Rules 25

Hazard Communication Program 10
Hazards 6
Health Information Privacy Act of 1996 54

IEEE Standard 13
Individual Responsibilities 38
Installation 75
Installation Details (Wiring Diagrams) 13
Instrument Index (List) 13
International Society of Automation 2
Interpersonal Communication Skills 21

Language 29
Lesson Learned Session 66
Location Plans 14
Lock out/tag out 8
Logic Diagrams 13

Material Safety Data Sheets (MSDS) 8
Monitoring Methods 43

NERC Critical Infrastructure Protection (CIP) 10

Obstacles to Effective Communication 29
Oral Skills 24
OSHA Hazard Communication Standard 29 CFR 1910.1200 8
Other Communications Concerns 28
Overall Project Coordination 38

Personnel 65
PERT Chart 37

Phase I – Initiation 71
Phase II – Planning 72
Phase III – Execution 74
Phase IV – Closeout 76
Physical Security Procedures 10
Piping and Instrumentation Drawings (P&ID's) 13
Planning Process 25
PMT Diversity Issues 29
Politics 30
Print Management 54
Print Redlining Process 56
Process Flow Diagrams (PFD's) 13
Project
    Administration 49
    Change Orders 44
    Closeout 61
    Closeout Report 67
    Documentation Requirements 5
    Engineering Documentation 12
    Management 1
    Management Software 37
    Monitoring & Control 43
    Planning 35
    Planning & Resource Coordination 33
    Report 43
    Status Report 44
    Update Meeting Minutes 44
Project Management Team (PMT) 34
Project Manager (PM) 34
Proper Use of Acronyms 29
Punch Lists 62

Redlining 55
Reporting 46
Request for Proposal (RFP) 73
Request for Quote (RFQ) 73

Index

Responsibilities 63
Risks 46

Safety Procedures 6
Schedule 46
Site Procedures 5
Specification Forms 13
Start-up 75
Supervision 39
System Integration Companies (SIC) 34

Task Assignments 38
Technical Team Lead 34
Termination 62
The control of hazardous energy 8
Time Management 39
Timekeeping Report 44
Transfer of Ownership 63
Typical Employee & Contractor Security Badges 11
Typical Project Status Report 46
Typical Site ERP Manual 7
Typical Site Procedures 5

United States Occupational Safety & Health 8
United States Department of Homeland Security 5, 10

Vendor Selection 73
Vendor/Supplier Status Reports 44

Who-Does What-By When (WWW) 25
Work Authorization Permit 8
Work Authorization Permit (WAP) 8
Work Orders 44
Writing Skills 22

NOTES

NOTES

NOTES

NOTES

Southern Wake Campus Library
Wake Tech Community College
9101 Fayetteville Road
Raleigh, NC 27603

WITHDRAWN

DATE DUE

8/18   PRINTED IN U.S.A.